MAGNETIC MATERIALS
Fundamentals and device applications

This book covers the fundamentals of magnetism and the basic theories and applications of conventional magnetic materials. In addition there is extensive discussion of novel magnetic phenomena and their modern device applications.

The book starts with a review of elementary magnetostatics and magnetic materials, followed by a discussion of the atomic origins of magnetism. The properties and applications of ferro-, ferri-, para-, dia- and antiferromagnets are surveyed, and the basic theories that describe them are outlined. The final part of the book focuses on novel magnetic phenomena, and on magnetic materials in modern technological applications.

Based on a course given by the author in the Materials Department at The University of California, Santa Barbara, the book is targeted at advanced undergraduate and graduate students as well as researchers new to the field. Highly illustrated, containing numerous homework exercises and worked solutions, this book is ideal for a one-semester course in magnetic materials.

NICOLA SPALDIN is an Associate Professor in the Materials Department at UC Santa Barbara, which she joined following a post-doctoral year in Applied Physics at Yale University and a PhD in Chemistry at UC Berkeley. She is an enthusiastic and accomplished teacher, with experiences ranging from tutoring in undergraduate residence halls, developing and managing the UCSB integrative graduate teaching program (IGERT), answering elementary school students' questions on-line, and teaching in rural high schools in Nepal. In 2001 her efforts were recognized with the award of the UCSB Distinguished Teaching Award. Her research is in the area of computational magnetic materials, including development of new methods and algorithms, developing fundamental understanding of material properties, and design of improved materials for device applications. Specific systems of interest include "multiferroic" materials, and magnetic semiconductor nanostructures. Recognitions for her research include an Alfred P. Sloan Foundation Fellowship in 2002, an Office of Naval Research Young Investigator Award and Technology Review Magazine's Young Innovator Award in 1999.

"Magnus magnes ipse est globus terrestris."
William Gilbert, *De magnete*. 1600.

MAGNETIC MATERIALS

Fundamentals and device applications

NICOLA A. SPALDIN

CAMBRIDGE
UNIVERSITY PRESS

PUBLISHED BY THE PRESS SYNDICATE OF THE UNIVERSITY OF CAMBRIDGE
The Pitt Building, Trumpington Street, Cambridge, United Kingdom

CAMBRIDGE UNIVERSITY PRESS
The Edinburgh Building, Cambridge CB2 2RU, UK
32 Avenue of the Americas, New York, NY 10013-2473, USA
477 Williamstown Road, Port Melbourne, VIC 3207, Australia
Ruiz de Alarcón 13, 28014 Madrid, Spain
Dock House, The Waterfront, Cape Town 8001, South Africa

http://www.cambridge.org
Information on this title: www.cambridge.org/9780521816311

First published 2003
Reprinted 2006

Printed in the United Kingdom at the University Press, Cambridge

Typeface Times 11/14 pt   *System* LaTeX 2_ε [TB]

A catalogue record for this book is available from the British Library

Library of Congress Cataloguing in Publication data

Spaldin, Nicola A. (Nicola Ann), 1969–
Magnetic materials : fundamentals and device applications / Nicola A. Spaldin
p.   cm.
Includes bibliographical references and index.
ISBN 0 521 81631 9 – ISBN 0 521 01658 4 (pb.)
1. Magnetic materials.   2. Magnetic devices.   3. Magnetism.   I. Title.
TK7872.M25 H54 2003
621.34 – dc21   2002073929

ISBN-13 978-0-521-81631-1 hardback
ISBN-10 0-521-81631-9 hardback

ISBN-13 978-0-521-01658-2 paperback
ISBN-10 0-521-01658-4 paperback

Contents

Acknowledgements

This book has been tested on human subjects during a course on Magnetic Materials that I have taught at UC Santa Barbara for the last few years. I am immensely grateful to each class of students for suggesting improvements, hunting for errors and letting me know when I am being boring. I hope that their enthusiasm is contagious.

Nicola Spaldin

1

Review of basic magnetostatics

"Mention magnetics and an image arises of musty physics labs peopled
by old codgers with iron filings under their fingernails."

John Simonds, *Physics Today* (April), 1995.

Before we can begin our discussion of magnetic materials we need to understand
some of the basic concepts of magnetism, such as what causes magnetic fields,
and what effects magnetic fields have on their surroundings. These fundamental
issues are the subject of this first chapter. Unfortunately, we are going to immedi-
ately run into a complication. There are two complementary ways of developing
the theory and definitions of magnetism. The 'physicist's way' is in terms of circulating currents, and the 'engineer's way' is in terms of magnetic poles (such as
we find at the ends of a bar magnet). The two developments lead to different views
of which interactions are more fundamental, to slightly different-looking equations, and (to really confuse things) to two different sets of units. Most books that
you'll read choose one convention or the other and stick with it. Instead, throughout this book we are going to follow what happens in 'real life' (or at least at
scientific conferences on magnetism) and use whichever convention is most appropriate to the particular problem. We'll see that it makes most sense to use Système
International d'Unités (SI) units when we talk in terms of circulating currents, and
centimeter–gram–second (cgs) units for describing interactions between magnetic
poles.

To avoid total confusion later, we will give our definitions in this chapter and the
next from *both* viewpoints, and provide a conversion chart for units and equations
at the end of Chapter 2. Reference 1 provides an excellent light-hearted discussion
of the unit systems used in describing magnetism.

1.1 Magnetic field

1.1.1 Magnetic poles

So let's begin by defining the magnetic field, **H**, in terms of magnetic poles. This is the order in which things happened historically – the law of interaction between magnetic poles was discovered by Michell in England in 1750, and by Coulomb in France in 1785, a few decades before magnetism was linked to the flow of electric current. These gentlemen found empirically that the force between two magnetic poles is proportional to the product of their pole strengths, p, and inversely proportional to the square of the distance between them,

$$F \propto \frac{p_1 p_2}{r^2}. \tag{1.1}$$

This is analogous to Coulomb's law for electric charges, with one important difference – scientists believe that single magnetic poles (magnetic monopoles) do not exist. They can however be approximated by one end of a very long bar magnet, which is how the experiments were carried out. By convention, the end of a freely suspended bar magnet which points towards magnetic north is called the north pole, and the opposite end is called the south pole.[†] In cgs units, the constant of proportionality is unity, so

$$F = \frac{p_1 p_2}{r^2} \qquad \text{(cgs)}, \tag{1.2}$$

where r is in centimeters and F is in dynes. Turning Eqn 1.2 around gives us the definition of pole strength:

A pole of unit strength is one which exerts a force of one dyne on another unit pole located at a distance of one centimeter.

The unit of pole strength does not have a name in the cgs system.

In SI units, the constant of proportionality in Eqn 1.1 is $1/4\pi\mu_0$, so

$$F = \frac{1}{4\pi\mu_0} \frac{p_1 p_2}{r^2} \qquad \text{(SI)}, \tag{1.3}$$

where μ_0 is called the permeability of free space, and has the value $4\pi \times 10^{-7}$ weber/ampere meter (Wb/Am). The SI unit of force is the newton (N), and 1 newton $= 10^5$ dyne (dyn).

To understand what causes the force, we can think of the first pole generating a magnetic field, **H**, which in turn exerts a force on the second pole. So

$$F = \left(\frac{p_1}{r^2}\right) p_2 = Hp_2 \tag{1.4}$$

[†] Note however that if we think of the earth's magnetic field as originating from a bar magnet, then the *south* pole of the earth's 'bar magnet' is actually at the magnetic north pole!

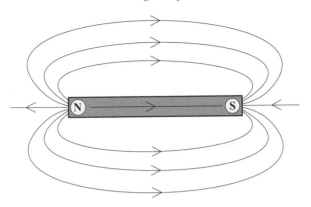

Figure 1.1 Field lines around a bar magnet. By convention, the lines originate at the north pole and end at the south pole.

giving, by definition,

$$H = \frac{p_1}{r^2}.$$ (1.5)

So:

A field of unit strength is one which exerts a force of one dyne on a unit pole.

By convention, the north pole is the *source* of the magnetic field, and the south pole is the *sink*, so we can sketch the magnetic field lines around a bar magnet as shown in Fig. 1.1.

The units of magnetic field are oersteds (Oe) in cgs units, so a field of unit strength has an intensity of one oersted. In the SI system, the field from a pole is

$$H = \frac{p}{4\pi \mu_0 r^2}$$ (1.6)

and the units are amperes per meter (A/m); $1\,\text{Oe} = (1000/4\pi)\,\text{A/m}$.

The earth's magnetic field has an intensity of around one-tenth of an oersted, and the field at the end of a typical kindergarten toy bar magnet is around 5000 Oe.

1.1.2 Magnetic flux

It's appropriate next to introduce another rather abstract concept, that of *magnetic flux*, Φ. The idea behind the term 'flux' is that the field of a magnetic pole is conveyed to a distant place by something which we call a flux. Rigorously the flux is defined as the surface integral of the normal component of the magnetic field. This means that the amount of flux passing through unit area perpendicular to the field is equal to the field strength. So the field strength is equal to the amount of flux per unit area, and the flux is the field strength *times* the area,

$$\Phi = \boldsymbol{H}A. \tag{1.7}$$

The unit of flux in cgs units, the oersted cm^2, is called the maxwell (Mx). In SI units the expression for flux is

$$\Phi = \mu_0 \boldsymbol{H}A \tag{1.8}$$

and the unit of flux is called the weber.

Magnetic flux is important because a *changing* flux generates an electric current in any circuit which it intersects. In fact we define an 'electromotive force', ε, equal to the rate of change of the flux linked with the circuit:

$$\varepsilon = -\frac{d\Phi}{dt}. \tag{1.9}$$

Equation 1.9 is Faraday's law of electromagnetic induction. The electromotive force provides the potential difference which drives electric current around the circuit. The minus sign in Eqn 1.9 shows us that the current sets up a magnetic field which acts in the opposite direction to the magnetic flux. (This is known as Lenz's law).[†]

The phenomenon of electromagnetic induction leads us to an alternative definition of flux, which is (in SI units):

A flux of one weber, when reduced to zero in one second, produces an electromotive force of one volt in a one-turn coil through which it passes.

1.1.3 Circulating currents

The next development in the history of magnetism took place in Denmark in 1820 when Oersted discovered that a magnetic compass needle is deflected in the neighborhood of an electric current. This was really a huge breakthrough because it unified two sciences. The new science of electromagnetism, which dealt with forces between moving charges and magnets, encompassed both electricity, which described the forces between charges, and magnetism, which described the forces between magnets.

Then Ampère discovered (again experimentally) that the magnetic field of a small current loop is identical to that of a small magnet. (By small we mean small with respect to the distance at which the magnetic field is observed). The north pole of a bar magnet corresponds to current circulating in a counter-clockwise direction, whereas clockwise current is equivalent to the south pole, as shown in Fig. 1.2. In addition, Ampère hypothesized that *all* magnetic effects are due to current loops, and that the magnetic effects in magnetic materials such as iron are due to so-called

---

[†] We won't cover electromagnetic induction in much detail in this book. A good introductory text is Ref. 2.

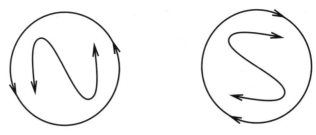

Figure 1.2 Relationship between direction of current flow and magnetic pole type.

'molecular currents'. This was remarkably insightful considering that the electron would not be discovered for another 100 years! Today it's believed that magnetic effects are caused by the orbital and spin angular momenta of electrons.

This leads us to an alternative definition of the magnetic field, in terms of current flow:

A current of one ampere passing through an infinitely long straight wire generates a field of strength $1/2\pi$ amperes per meter at a radial distance of one meter.

Of course the next obvious question to ask is what happens if the wire is *not* straight? What magnetic field does a *general* circuit produce? Ampère solved this one too.

1.1.4 Ampère's circuital law

Ampère observed that the magnetic field generated by an electrical circuit depends on both the *shape* of the circuit *and* on the amount of current being carried. In fact the total current, I, is equal to the line integral of the magnetic field around a closed path containing the current. In SI units

$$\oint \boldsymbol{H}\cdot d\boldsymbol{l} = I. \tag{1.10}$$

This expression is called Ampère's circuital law, and it can be used to calculate the field produced by a current-carrying conductor. We will look at some examples later.

1.1.5 Biot–Savart law

An equivalent statement to Ampère's circuital law (which is sometimes easier to use for particular symmetries) is given by the Biot–Savart law. The Biot–Savart law gives the field contribution, $\delta\boldsymbol{H}$, generated by a current flowing in an elemental

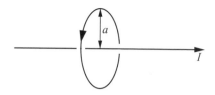

Figure 1.3 Calculation of the field from a current flowing in a long straight wire, using Ampère's circuital law.

length, δl of a conductor:

$$\delta H = \frac{1}{4\pi r^2} I \delta l \times \hat{u} \qquad (1.11)$$

where r is the radial distance from the conductor, and $\hat{u}$ is a unit vector along the radial direction.

1.1.6 Field from a straight wire

To show that these laws are equivalent, let's use them both to calculate the magnetic field generated by a current flowing in a straight wire.

First using Ampère's law. The geometry of the problem is shown in Fig. 1.3. If we assume that the field lines go around the wire in closed circles (by symmetry this is a fairly safe assumption) then the field, H, has the same value at all points on a circle concentric with the wire. This makes the line integral of Eqn 1.10 straightforward. It's just

$$\oint H \cdot dl = 2\pi a H = I \qquad \text{by Ampère's law,} \qquad (1.12)$$

and so the field, H, at a distance a from the wire is

$$H = \frac{I}{2\pi a}. \qquad (1.13)$$

For this particular problem, the Biot–Savart law is somewhat less straightforward to apply. The geometry for calculating the field at a point P at a distance a from the wire is shown in Fig. 1.4. Now

$$\begin{aligned} \delta H &= \frac{1}{4\pi r^2} I \delta l \times \hat{u} \\ &= \frac{1}{4\pi r^2} I |\delta l| |\hat{u}| \sin\theta \end{aligned} \qquad (1.14)$$

where θ is the angle between δl and $\hat{u}$, which is equal to $(90 + \alpha)$. So

$$\begin{aligned} \delta H &= \frac{I}{4\pi r^2} \delta l \sin(90 + \alpha) \\ &= \frac{I}{4\pi r^2} \frac{r \delta \alpha}{\cos \alpha} \sin(90 + \alpha) \end{aligned} \qquad (1.15)$$

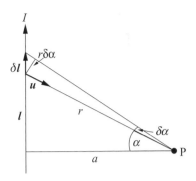

Figure 1.4 Calculation of the field from a current flowing in a long straight wire, using the Biot–Savart law.

since $\delta l = r\delta\alpha/\cos\alpha$.

But $\sin(90 + \alpha) = \cos\alpha$, and $r = a/\cos\alpha$. So

$$\delta H = \frac{I}{4\pi}\frac{\cos^2\alpha}{a^2}\frac{a\delta\alpha}{\cos^2\alpha}\cos\alpha$$
$$= \frac{I\cos\alpha\,\delta\alpha}{4\pi a} \tag{1.16}$$

and

$$H = \frac{I}{4\pi a}\int_{-\pi/2}^{\pi/2}\cos\alpha\,d\alpha$$
$$= \frac{I}{4\pi a}[\sin\alpha]_{-\pi/2}^{\pi/2}$$
$$= \frac{I}{2\pi a}. \tag{1.17}$$

The same result as that obtained using Ampère's law! Clearly Ampère's law was a better choice for this particular problem.

Unfortunately, analytic expressions for the field produced by a current can only be obtained for conductors with rather simple geometries. For more complicated shapes the field must be calculated numerically. Numerical calculation of magnetic fields is an active research area, and is tremendously important in the design of electromagnetic devices. A review is given in Ref. 3.

1.2 Magnetic moment

Next we need to introduce the concept of magnetic moment, which is the moment of the couple exerted on either a bar magnet or a current loop when it is in an applied field. Again we can define the magnetic moment either in terms of poles, or in terms of currents.

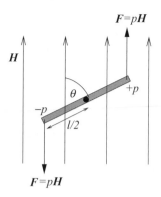

Figure 1.5 Calculation of the moment exerted on a bar magnet in a magnetic field. From *Introduction to magnetic materials* by Cullity, © Reprinted by permission of Pearson Education, Inc., Upper Saddle River, NJ.

Imagine a bar magnet is at an angle θ to a magnetic field, $\boldsymbol{H}$, as shown in Fig. 1.5. We showed in Section 1.1.1 that the force on each pole, $\boldsymbol{F} = p\boldsymbol{H}$. So the moment acting on the magnet, which is just the force times the perpendicular distance from the center of mass, is

$$pH \sin\theta \frac{l}{2} + pH \sin\theta \frac{l}{2} = pHl \sin\theta = mH \sin\theta \qquad (1.18)$$

where $\boldsymbol{m} = pl$, the product of the pole strength and the length of the magnet, is the *magnetic moment*. (Our notation here is to represent vector quantities by bold italic type, and their magnitudes by regular italic type). This gives a definition:

The magnetic moment is the moment of the couple exerted on a magnet when it is perpendicular to a uniform field of 1 oersted.

Alternatively, if a current loop has area A and carries a current I, then its magnetic moment is defined as

$$\boldsymbol{m} = IA. \qquad (1.19)$$

The cgs unit of magnetic moment is the emu. In SI units, magnetic moment is measured in A m^2.

1.2.1 Magnetic dipole

A magnetic dipole is defined as either the magnetic moment, $\boldsymbol{m}$, of a bar magnet in the limit of small *length* but finite moment, or the magnetic moment $\boldsymbol{m}$ of a current loop in the limit of small *area* but finite moment. The field lines around a magnetic dipole are shown in Fig. 1.6. The energy of a magnetic dipole is defined to be zero when the dipole is perpendicular to a magnetic field. So the work done (in ergs) in

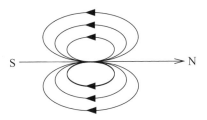

Figure 1.6 Field lines around a magnetic dipole.

turning through an angle $d\theta$ against the field is

$$dE = 2(pH \sin\theta)\frac{l}{2}d\theta$$
$$= mH \sin\theta \, d\theta \qquad (1.20)$$

and the energy of a dipole at an angle θ to a magnetic field is

$$E = \int_{\pi/2}^{\theta} mH \sin\theta \, d\theta$$
$$= -mH \cos\theta$$
$$= -\boldsymbol{m} \cdot \boldsymbol{H}. \qquad (1.21)$$

This expression for the energy of a magnetic dipole in a magnetic field is in cgs units. In SI units the energy is $E = -\mu_0 \boldsymbol{m} \cdot \boldsymbol{H}$. We will be using the concept of magnetic dipole, and this expression for its energy in a magnetic field, extensively throughout this book.

1.3 Definitions

Finally for this chapter, let's review the definitions which we've introduced so far. Here we give all the definitions in cgs units.

1. **Magnetic pole, p.** A pole of unit strength is one which exerts a force of one dyne on another unit pole located at a distance of one centimeter.
2. **Magnetic field, H.** A field of unit strength is one which exerts a force of one dyne on a unit pole.
3. **Magnetic flux, Φ.** The amount of magnetic flux passing through an area A is equal to the product of the magnetic field strength and the area; $\Phi = HA$.
4. **Magnetic moment, m.** The magnetic moment of a magnet is the moment of the couple exerted on the magnet when it is perpendicular to a uniform field of one oersted. For a bar magnet, $m = pl$, where p is the pole strength, and l is the length of the magnet.
5. **Magnetic dipole.** The energy of a magnetic dipole in a magnetic field is the dot product of the magnetic moment and the magnetic field; $E = -\boldsymbol{m} \cdot \boldsymbol{H}$.

Homework

Exercises

1.1   Using either the Biot–Savart law or Ampère's circuital law, derive a general expression for the magnetic field produced by a current flowing in a circular coil, at the center of the coil.

1.2a   Derive an expression for the field produced by a current flowing in a circular coil, at a general point on the *axis* of the coil.

1.2b   Could we derive a corresponding analytic expression for the field at a general, off-axis point? If not, how might we go about calculating magnetic fields for generalized geometries?

1.3a   Calculate the field generated by an electron moving in a circular orbit of radius 1 Å (1 Å $= 10^{-10}$ m) with angular momentum $\hbar$ J s, at a distance of 3 Å from the center of the orbit, and along its axis.

1.3b   Calculate the magnetic dipole moment of the electron in 1.3a. Give your answer in SI and cgs units.

1.3c   Calculate the magnetic dipolar energy of the circulating electron in 1.3a, when it is in the field generated by a second identical circulating electron at a distance of 3 Å away along its axis. Assume that the magnetic moment of the first electron is aligned parallel to the field from the second electron.

1.4   Derive an expression for the field H produced by 'Helmholtz coils', that is, two co-axial coils each of radius a, and separated by a distance a, at a point on the axis x between the coils:

  (a) with current flowing in the same sense in each coil, and

  (b) with current flowing in the opposite sense in each coil. In this case, derive the expression for dH/dx also.

For $a = 1$ m, and for both current orientations, calculate the value of the field halfway between the coils, and at $\frac{1}{4}$ and $\frac{3}{4}$ along the axis. What qualitative feature of the field is significant in each case? Suggest a use for each pair of Helmholtz coils.

Further reading

D. Jiles, *Introduction to magnetism and magnetic materials*. Chapman & Hall, 1996, Chapter 1.

B.D. Cullity, *Introduction to magnetic materials*. Addison-Wesley, 1972, Chapter 1.

2

Magnetization and magnetic materials

"Modern technology would be unthinkable without magnetic materials
and magnetic phenomena."
Rolf E. Hummel, *Understanding materials science.* Springer, 1998.

Now that we have covered some of the fundamentals of magnetism, we are allowed
to start on the fun stuff! In this chapter we will learn about the magnetic field *inside*
materials, which is generally quite different from the magnetic field *outside*. Most
of the technology of magnetic materials is based on this simple statement, and this
is why the study of magnetism is exciting for materials scientists.

2.1 Magnetic induction and magnetization

When a magnetic field, H, is applied to a material, the response of the material is
called its *magnetic induction*, B. The relationship between B and H is a property
of the material. In some materials (and in free space) B is a linear function of H but
in general it is much more complicated, and sometimes it's not even single-valued.
The equation relating B and H is (in cgs units)

$$B = H + 4\pi M, \tag{2.1}$$

where M is the *magnetization* of the medium. The magnetization is defined to be
the magnetic moment per unit volume,

$$M = \frac{m}{V} \quad \frac{\text{emu}}{\text{cm}^3}. \tag{2.2}$$

M is a property of the material, and depends on both the individual magnetic
moments of the constituent ions, atoms or molecules, and on how these dipole
moments interact with each other. The cgs unit of magnetization is the emu/cm^3,
and that of magnetic induction is the gauss.

11

In SI units the relationship between B, H and M is

$$B = \mu_0(H + M),\qquad(2.3)$$

where μ_0 is the permeability of free space. The units of M are obviously the same as those of H (A/m), and those of μ_0 are weber/A m, also known as henry/m. So the units of B are weber/m², or tesla (T); 1 gauss $= 10^{-4}$ tesla.

2.2 Flux density

The magnetic induction, B, is the same thing as the density of flux, Φ, *inside* the medium. So within a material, $B = \Phi/A$, by analogy with $H = \Phi/A$ in free space. In general the flux density inside a material is different from that outside. In fact magnetic materials can be classified based on the difference between their internal and external flux.

If Φ inside is *less than* Φ outside then the material is known as *diamagnetic*. Examples of diamagnetic materials include bismuth and helium. These materials tend to exclude the magnetic field from their interior. We'll see later that the atoms or ions which make up diamagnetic materials have zero magnetic dipole moment. If Φ inside is *slightly more* than Φ outside then the material is either paramagnetic (e.g. Na or Al) or antiferromagnetic (e.g. MnO or FeO). In many paramagnetic and antiferromagnetic materials, the constituent atoms or ions have a magnetic dipole moment. In paramagnets these dipole moments are randomly oriented, and in antiferromagnets they are ordered antiparallel to each other. Finally, if Φ inside is *very much greater* than Φ outside then the material is either ferromagnetic or ferrimagnetic. In ferromagnets, the magnetic dipole moments of the atoms tend to line up in the same direction. Ferrimagnets are somewhat like antiferromagnets, in that the dipoles align antiparallel, however some of the dipole moments are larger than others, so the material has a net overall magnetic moment. Ferro- and ferrimagnets tend to concentrate magnetic flux in their interiors. Figure 2.1 shows these different kinds of magnetic materials schematically. The reason for the different types of ordering, and the resulting material properties, are the subjects of much of the rest of this book.

2.3 Susceptibility and permeability

The properties of a material are defined not only by the magnetization, or the magnetic induction, but by the way in which these quantities *vary* with the applied magnetic field.

The ratio of M to H is called the susceptibility

$$\chi = \frac{M}{H}\qquad \frac{\text{emu}}{\text{cm}^3\,\text{Oe}}.\qquad(2.4)$$

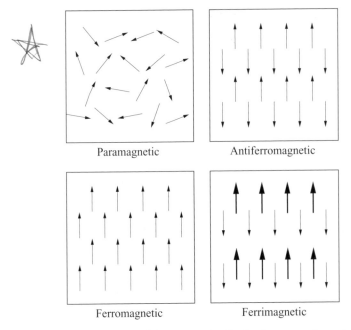

Figure 2.1 Ordering of the magnetic dipoles in magnetic materials.

The susceptibility indicates how responsive a material is to an applied magnetic field. (Sometimes the symbol κ is used for the susceptibility per unit volume, then $\chi = \kappa/\rho$ emu/g Oe is the susceptibility per unit mass.)

The ratio of $\boldsymbol{B}$ to $\boldsymbol{H}$ is called the permeability

$$\mu = \frac{\boldsymbol{B}}{\boldsymbol{H}} \quad \frac{\text{gauss}}{\text{Oe}}, \tag{2.5}$$

and μ indicates how *permeable* the material is to the magnetic field. A material which concentrates a large amount of flux density in its interior has a high permeability. Using the relationship $\boldsymbol{B} = \boldsymbol{H} + 4\pi\boldsymbol{M}$ gives us the relationship (in cgs units) between the permeability and susceptibility:

$$\mu = 1 + 4\pi\chi. \tag{2.6}$$

Note that in SI units, the susceptibility is dimensionless, and the permeability is in units of henry/m. The corresponding relationship between permeability and susceptibility in SI units is

$$\frac{\mu}{\mu_0} = 1 + \chi, \tag{2.7}$$

where μ_0 (see Eqn 1.3) is the permeability in free space.

Graphs of $\boldsymbol{M}$ or $\boldsymbol{B}$ versus $\boldsymbol{H}$ are called *magnetization curves*, and are characteristic of the type of material. Let's look at a few, for the most common types of magnetic materials.

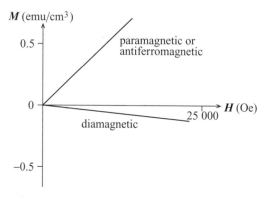

Figure 2.2 Magnetization curves for dia-, para- and antiferromagnets.

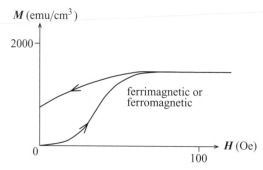

Figure 2.3 Magnetization curves for ferri- and ferromagnets.

The magnetization of dia-, para- and antiferromagnetic materials is plotted schematically as a function of applied field in Fig. 2.2. For all these materials the *M–H* curves are linear. Rather large applied fields are required to cause rather small changes in magnetization, and no magnetization is retained when the applied field is removed. For diamagnets, the slope of the *M–H* curve is negative, so the susceptibility is small and negative, and the permeability is slightly less than one. For para- and antiferromagnets the slope is positive and the susceptibility and permeability are correspondingly small and positive, and slightly greater than unity respectively.

Figure 2.3 shows schematic magnetization curves for ferri- and ferromagnets. The first point to note is that the axis scales are completely different from those in Fig. 2.2. In this case, a much larger magnetization is obtained on application of a much smaller external field. Second, the magnetization *saturates* – above a certain applied field, an increase in field causes only a very small increase in magnetization. Clearly both χ and μ are large and positive, and are functions of the applied field. Finally, decreasing the field to zero after saturation does not reduce the magnetization to zero. This phenomenon is called *hysteresis*, and is very important in technological applications. For example the fact that ferro- and ferrimagnetic materials retain their magnetization in the absence of a field allows them to be made into permanent magnets.

2.4 Hysteresis loops

We've just seen that reducing the field to zero does not reduce the magnetization of a ferromagnet to zero. In fact ferro- and ferrimagnets continue to show interesting behavior when the field is reduced to zero and then reversed in direction. The graph of B (or M) versus H which is traced out is called a *hysteresis loop*. Figure 2.4 shows a schematic of a generic hysteresis loop – this time we've plotted B versus H.

Our magnetic material starts at the origin in an unmagnetized state, and the magnetic induction follows the curve from 0 to B_s as the field is increased in the positive direction. Note that, although the magnetization is constant after saturation (as we saw in Fig. 2.3), B continues to increase, because $B = H + 4\pi M$. The value of B at B_s is called the saturation induction, and the curve of B from the demagnetized state to B_s is called the *normal induction* curve.

When H is reduced to zero *after* saturation, the induction decreases from B_s to B_r – the *residual induction*, or retentivity. The reversed field required to reduce the induction to zero is called the *coercivity*, H_c. Depending on the value of the coercivity, ferromagnetic materials are classified as either hard or soft. A hard magnet needs a large field to reduce its induction to zero (or conversely to saturate the magnetization). A soft magnet is easily saturated, but also easily demagnetized. Hard and soft magnetic materials obviously have totally complementary applications!

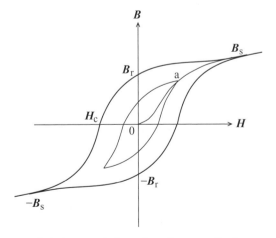

Figure 2.4 Hysteresis loop for a ferro- or ferrimagnet.

When the reversed H is increased further, saturation is achieved in the reverse direction. The loop that is traced out is called the *major hysteresis* loop. Both tips represent magnetic saturation, and there is inversion symmetry about the origin. If the initial magnetization is interrupted (for example at point a), and H is reversed then re-applied, then the induction follows a *minor* hysteresis loop.

The suitability of ferri- and ferromagnetic materials for particular applications is determined largely from characteristics shown by their hysteresis loops. We'll discuss the origin of hysteresis, and the relationship between hysteresis loop and material properties, in the later chapters devoted to ferromagnetic and ferrimagnetic materials.

2.5 Definitions

Let's review the new definitions which we have introduced in this chapter.

1. **Magnetic induction, B**. The magnetic induction is the response of a material to a magnetic field, H.
2. **Magnetization, M**. The magnetization is the total magnetic moment per unit volume.
3. **Susceptibility, χ**. The susceptibility is the ratio of M to H.
4. **Permeability, μ**. The permeability is the ratio of B to H.

2.6 Units and conversions

Finally for this chapter we provide a conversion chart between cgs and SI for the units and equations which we have introduced so far.

Equation conversions

	cgs	SI
force between poles	$F = \dfrac{p_1 p_2}{r^2}$ (dyne)	$F = \dfrac{1}{4\pi\mu_0}\dfrac{p_1 p_2}{r^2}$ (newton)
field of a pole	$H = \dfrac{p}{r^2}$ (oersted)	$H = \dfrac{1}{4\pi\mu_0}\dfrac{p}{r^2}$ (ampere/m)
magnetic induction	$B = H + 4\pi M$ (gauss)	$B = \mu_0(H + M)$ (tesla)
energy of a dipole	$E = -\mathbf{m}\cdot\mathbf{H}$ (erg)	$E = -\mu_0\mathbf{m}\cdot\mathbf{H}$ (joule)
susceptibility	$\chi = \dfrac{M}{H}$ (emu/cm³ oersted)	$\chi = \dfrac{M}{H}$ (dimensionless)
permeability	$\mu = \dfrac{B}{H} = 1 + 4\pi\chi \left(\dfrac{\text{gauss}}{\text{oersted}}\right)$	$\mu = \dfrac{B}{H} = \mu_0(1 + \chi) \left(\dfrac{\text{henry}}{\text{m}}\right)$

Unit conversions

F	1 dyne	=	10^{-5} newton
H	1 oersted	=	79.58 ampere/m
B	1 gauss	=	10^{-4} tesla
E	1 erg	=	10^{-7} joule
Φ	1 maxwell	=	10^{-8} weber
M	1 emu/cm^3	=	12.57×10^{-4} weber/m^2
μ	1 gauss/oersted	=	1.257×10^{-6} henry/m

It is often useful to convert the SI units into their fundamental constituents, ampere (A), meter (m), kilogram (kg) and second (s). Here are some examples.

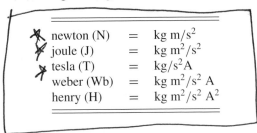

newton (N)	=	kg m/s^2
joule (J)	=	kg m^2/s^2
tesla (T)	=	kg/s^{2}A
weber (Wb)	=	kg m^2/s^2 A
henry (H)	=	kg m^2/s^2 A^2

Homework

Exercises

2.1 A cylindrical bar magnet 10 inches long and 1 inch in diameter has a magnetic moment of 10 000 erg/Oe.

(a) What is its magnetic moment in SI units?

(b) What is its magnetization in both cgs and SI units?

(c) What current would have to be passed through a 100-turn solenoid of the same dimensions to give it the same magnetic moment?

Further reading

IEEE Trans. Mag. **20**, 112, 1984.
D. Jiles, *Introduction to magnetism and magnetic materials*. Chapman & Hall, 1996, Chapter 2.

To think about

What does it mean to have a permeability of zero? What is the corresponding value of susceptibility? Can you think of any materials that have these properties?

3

Atomic origins of magnetism

"Only in a few cases have results of direct chemical interest been obtained
by the accurate solution of the Schrödinger equation."
Linus Pauling, *The nature of the chemical bond.* Cornell University
Press, 1960.

The purpose of this chapter is to understand the origin of the magnetic dipole
moment of free atoms. We will make the link between Ampère's ideas about circulating currents, and the electronic structure of atoms. We'll see that it is the angular
momentum of the electrons in atoms which correspond to Ampère's circulating
currents and give rise to the magnetic dipole moment.

In fact we will see that the magnetic moment of a free atom in the absence of
a magnetic field consists of *two* contributions. First the orbital angular momenta
of the electrons circulating the nucleus. In addition each electron has an extra
contribution to its magnetic moment arising from its 'spin'. The spin and orbital
angular momenta combine to produce the observed magnetic moment.[†]

By the end of this chapter we will understand some of the quantum mechanics which explains why some isolated atoms have a permanent magnetic dipole
moment, and others do not. And we will develop some rules for determining the
magnitudes of these dipole moments. Later in the book we will look at what happens
to these dipole moments when we combine the atoms into molecules and solids.

3.1 Solution of the Schrödinger equation for a free atom

We begin with a review of atomic theory to show how solution of the Schrödinger
equation leads to *quantization* of the orbital angular momentum of the electrons.

[†] In the presence of an external field there is a third contribution to the magnetic moment of a free atom arising from
the change in orbital angular momentum due to the applied field. We will investigate this further in Chapter 4
when we discuss diamagnetism.

The quantization is important because it means that the atomic dipole moments are restricted to certain values and to certain orientations with respect to an external field. We'll see later that these restrictions have a profound effect on the properties of magnetic materials.

For simplicity we'll consider the hydrogen atom, which consists of a single negatively charged electron bound to a positively charged nucleus. The potential energy of the hydrogen atom is just the Coulomb interaction between the electron and the nucleus, $-e^2/4\pi\epsilon_0 r$ where e is the charge on the electron and ϵ_0 is the permittivity of free space. So the Schrödinger equation, $H\Psi = E\Psi$, becomes

$$-\frac{\hbar^2}{2m_e}\nabla^2\Psi - \frac{e^2}{4\pi\epsilon_0 r}\Psi = E\Psi, \qquad (3.1)$$

where m_e is the mass of the electron and

$$\nabla^2 = \frac{1}{r}\frac{\partial^2}{\partial r^2}r + \frac{1}{r^2}\left[\frac{1}{\sin^2\theta}\frac{\partial^2}{\partial\theta^2} + \frac{1}{\sin\theta}\frac{\partial}{\partial\theta}\sin\theta\frac{\partial}{\partial\theta}\right] \qquad (3.2)$$

(in spherical coordinates).

For bound states (with energy, E, less than zero) this has the well-known solution

$$\Psi_{nlm_l}(r, \theta, \phi) = R_{nl}(r)Y_{lm_l}(\theta, \phi). \qquad (3.3)$$

(You can find a complete derivation in most quantum mechanics text books – my personal favorite is in the *Feynman lectures on physics*, Ref. 4.) As a result of the spherical symmetry of the Coulomb potential, the wavefunction separates into a product of radial and angular functions! Physically, each wavefunction describes the radial and angular distributions of an electron with a particular set of n, l and m_l values. In particular, the probability of finding an electron in some infinitesimal region at the point r is given by $|\Psi_{nlm_l}(r, \theta, \phi)|^2$.

The radial part of the wavefunction is made up of the associated Laguerre functions, $R_{nl}(r)$, which are each specified by two labels, n and l. The first few Laguerre functions are tabulated in Table 3.1. The radial parts of the hydrogen atom wavefunctions with $l = 0$ (the s orbitals) and $n = 1, 2$ and 3, and $l = 1$ (the p orbitals) and $n = 2$ and 3, are plotted in Fig. 3.1. We see that as n increases the wavefunctions extend further from the nucleus. Also, the number of times the wavefunction crosses the zero axis (the number of nodes in the wavefunction) is equal to $n - l - 1$.

The spherical harmonics, $Y_{lm_l}(\theta, \phi)$, labeled by l and m_l, form the angular part of the wavefunctions. The first few spherical harmonics are tabulated in Table 3.2.

The n, l and m_l labels are *quantum numbers*, and they determine the form of the allowed solutions to the Schrödinger equation for the hydrogen atom. The n and l labels are called the *principal* and *angular momentum* quantum numbers respectively, and the label m_l is called the *magnetic* quantum number. The lowest

Table 3.1 *Radial dependence of the hydrogen atomic orbitals.*

n	l	$R_{nl}(r)$
1	0	$\left(\dfrac{1}{a_0}\right)^{3/2} 2e^{-r/a_0}$
2	0	$\left(\dfrac{1}{a_0}\right)^{3/2} \dfrac{1}{2\sqrt{2}}\left(2 - \dfrac{r}{a_0}\right) e^{-r/2a_0}$
2	1	$\left(\dfrac{1}{a_0}\right)^{3/2} \dfrac{1}{2\sqrt{6}}\dfrac{r}{a_0} e^{-r/2a_0}$

Table 3.2 *Angular dependence of the hydrogen atomic orbitals.*

l	m_l	$Y_{lm_l}(\theta, \phi)$
0	0	$\left(\dfrac{1}{2\pi}\right)^{1/2}$
1	0	$\dfrac{1}{2}\left(\dfrac{1}{3\pi}\right)^{1/2} \cos\theta$
1	1	$-\dfrac{1}{2}\left(\dfrac{1}{3\pi}\right)^{1/2} \sin\theta\, e^{+i\phi}$
1	-1	$+\dfrac{1}{2}\left(\dfrac{1}{3\pi}\right)^{1/2} \sin\theta\, e^{-i\phi}$

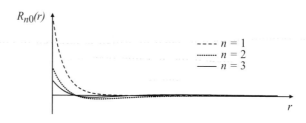

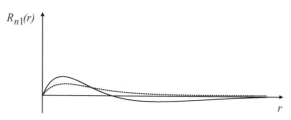

Figure 3.1 Radial parts of the hydrogen atom wavefunctions with $l = 0$ and $l = 1$.

energy configuration for the hydrogen atom has the electron in the $n = 1, l = 0$ level.

3.1.1 What do the quantum numbers represent?

The permitted solutions to the Schrödinger equation are restricted because the quantum numbers are only allowed to take the following values:

$$n = 1, 2, 3, \ldots \tag{3.4}$$
$$l = 0, 1, 2, \ldots, n - 1 \tag{3.5}$$
$$m_l = -l, -l + 1, \ldots, l - 1, l. \tag{3.6}$$

This means that the electron can only have certain radial and angular distributions, determined by it's quantum numbers. The quantum numbers in turn determine other properties of the electron.

The principal quantum number, n

The principal quantum number, n, determines the *energy*, E_n, of the electron level. (You might remember the n label from discussions of the Bohr atom in elementary atomic theory texts.) In the hydrogen atom the energy is given by

$$E_n = - \left(\frac{m_e e^4}{32\pi^2 \epsilon_0^2 \hbar^2} \right) \frac{1}{n^2}, \tag{3.7}$$

where $\hbar = h/2\pi$ is Planck's constant. Levels with smaller values of n (with $n = 1$ being the smallest that is allowed) have lower energy. Therefore in the ground state of the hydrogen atom, the single electron occupies the $n = 1$ energy level. Electrons with a particular n value are said to form the nth electron 'shell'. There are n^2 electronic orbitals in shell n, each of which is allowed to contain a maximum of two electrons. Although the n value does not directly determine any magnetic properties, we'll see later that it influences the magnetic properties of an atom because it controls which values of the l and m_l quantum numbers are permitted.

The orbital quantum number, l

The orbital quantum number, l, determines the magnitude of the orbital angular momentum of the electron. The *magnitude* of the orbital angular momentum, $|L|$, of an *individual* electron is related to the angular momentum quantum number, l, by

$$|L| = \sqrt{l(l + 1)}\hbar. \tag{3.8}$$

(We won't derive this result here – it comes from the fact that the spherical har-
monics satisfy the equation $\nabla^2 Y_{lm_l}(\theta, \phi) = -l(l+1)Y_{lm_l}(\theta, \phi)$. Again, Ref. 4 has
an excellent derivation.)

Values of l equal to 0, 1, 2, 3, etc., correspond respectively to the familiar labels
s, p, d and f for the atomic orbitals. (The labels s, p, d and f are legacies from old
spectroscopic observations of sharp, principal, diffuse and fundamental series of
lines.) We see that the s orbitals, with $l = 0$, and consequently $|L| = 0$, have zero
orbital angular momentum. So the electrons in s orbitals make no contribution to the
magnetic dipole of an atom from their orbital angular momentum. Similarly the p
electrons, with $l = 1$, have an orbital angular momentum of magnitude $|L| = \sqrt{2}\hbar$,
and so on for the orbitals of higher angular momentum.

The value of the angular momentum quantum number affects the radial distribution of the wavefunction, as we saw in Fig. 3.1. The s electrons, with $l = 0$,
have non-zero values at the nucleus, whereas the p electrons, with $l = 1$, have zero
probability of being found at the nucleus. We can think of this as resulting from the
orbital angular momentum's centrifugal force flinging the electron away from the
nucleus.

Since l can take integral values from 0 to $n - 1$, the $n = 1$ level contains only
s orbitals, the $n = 2$ level contains s and p, and the $n = 3$ level contains s, p and
d orbitals. Here we see the value of the n quantum number influencing the allowed
angular momentum of the electron. In our treatment of the hydrogen atom, all s,
p, d, etc., orbitals with the same n value have the same energy. We'll see later that
this is not the case in atoms with more than one electron, because the interactions
between the electrons affect the relative energies of states with different angular
momentum.

The magnetic quantum number, m_l

The *orientation* of the orbital angular momentum with respect to a magnetic field
is also quantized, and is labeled by the magnetic quantum number, m_l, which is
allowed to take integer values from $-l$ to $+l$. So (for example) a p orbital, with
$l = 1$, can have m_l values of -1, 0 or $+1$. This means that p orbitals can exist with
three orientations relative to an externally applied magnetic field.

The component of angular momentum along the field direction is equal to $m_l\hbar$.
For a p orbital this gives components of $+\hbar$, 0, or $-\hbar$, as illustrated in Fig. 3.2.
So the component of orbital angular momentum along the field direction is always
smaller than the total orbital angular momentum. (Remember the *magnitude* of the
orbital angular momentum for a p orbital is $\sqrt{l(l+1)}\hbar = \sqrt{2}\hbar$.) This means that
the orbital angular momentum vector can never point directly along the direction
of the field, and instead it *precesses* in a cone around the field direction, like a
gyroscope tipped off its axis. The cones of precession are shown schematically on

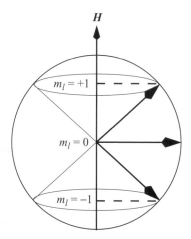

Figure 3.2 Component of angular momentum along the magnetic field direction for a p orbital (with $l = 1$). The radius of the circle is $\sqrt{2}\hbar$.

the figure in narrow lines. This off-axis precession is an intrinsic feature of the quantum mechanics of angular momentum – only in macroscopic objects, such as a spinning top, is the value of $\sqrt{l(l+1)}$ so close to l that the object appears to be able to rotate directly around the z axis.

For all three p orbitals, the component of angular momentum perpendicular to the applied field averages to zero.

3.2 Extension to many-electron atoms

In our hydrogen atom example, we saw that the wavefunction of an electron was completely separable into radial and angular parts, and that the energy of an electron depended only on the principal quantum number, n. In all atoms *other* than hydrogen, there is more than one electron, and the electrons interact with each other as well as with the nucleus. The additional interactions produce a much more complicated Schrödinger equation which can no longer be solved analytically.

One result of this 'many-body' effect is that the energy of an electron depends on both n and l. It is found that electrons with lower angular momentum (i.e. smaller l) are more stable. This leads to the familiar ordering of atomic orbitals through the periodic table:

$$1s; 2s, 2p; 3s, 3p, 3d; 4s, \ldots . \tag{3.9}$$

The ordering can be understood qualitatively by assuming that the electrons shield each other from the nucleus, thus reducing the energetically favorable Coulomb attraction. Wavefunctions with smaller l penetrate closer to the nucleus, and therefore have less shielding than those with higher l values. As a consequence they have lower energy.

In general, electrons fill the atomic orbitals in order, starting with those of lowest energy. So the 1s orbital is filled first, followed by the 2s, then the three 2p orbitals, etc. We will discuss more detailed rules for arranging the electrons so that they have the minimum energy within a particular set of n and l values later.

3.3 The normal Zeeman effect

We can see direct evidence that electrons behave like charged particles having angular momentum around the atomic nucleus, by observing the change in the atomic absorption spectrum in the presence of an external magnetic field.

In Chapter 1 we saw that the energy of a magnetic dipole moment, $\boldsymbol{m}$, in a magnetic field, $\boldsymbol{H}$, is given by

$$E = -\mu_0 \boldsymbol{m} \cdot \boldsymbol{H} \tag{3.10}$$

(in SI units).

We also showed that the magnetic dipole moment of a circulating current is given by

$$\boldsymbol{m} = I A, \tag{3.11}$$

where I is the current and A is the area of the circulating current loop.

By definition, the current, I, is just the charge passing per unit time. If we assume that the current is produced entirely by an electron orbiting at a distance a from the nucleus in an atom, then the magnitude of that current is equal to the charge on the electron multiplied by its velocity, v, divided by the circumference of the orbit ($2\pi a$):

$$I = \frac{ev}{2\pi a} = -\frac{|e|v}{2\pi a}. \tag{3.12}$$

Here the minus sign occurs because the charge on the electron is negative, and so the direction of current flow is opposite to that of the electron motion.

The area of the orbit is $A = \pi a^2$, so the magnetic dipole moment,

$$\boldsymbol{m} = I A = \frac{eva}{2} = -\frac{|e|va}{2}. \tag{3.13}$$

But the angular momentum of *any* object going round in a circle is the mass times the velocity times the distance from the axis ($=m_e v a$ in our case). And we stated in Section 3.1 that the orbital angular momentum about the magnetic field axis is only allowed to take the values $m_l \hbar$. So the angular momentum projected onto the field axis is

$$m_e va = m_l \hbar \tag{3.14}$$

giving

$$v = \frac{m_l \hbar}{m_e a}. \tag{3.15}$$

So, substituting for v in Eqn 3.10 gives the energy

$$E = \mu_0 \frac{e\hbar}{2m_e} m_l H = \mu_0 \mu_B m_l H. \tag{3.16}$$

(The corresponding expression in cgs units is $E = \mu_B m_l H$). The quantity $\mu_B = e\hbar/2m_e$ is called the Bohr magneton, and is the elementary unit of orbital magnetic moment in an atom. Its value is 9.274×10^{-24} J/T. (In cgs units it is written as $\mu_B = e\hbar/2m_e c = 0.927 \times 10^{-20}$ erg/Oe, where c is the velocity of light). So we see that the energy of an atomic orbital changes in the presence of a magnetic field, by an amount proportional to the orbital angular momentum of the orbital and the applied field strength. This phenomenon is known as the normal Zeeman effect[5] and can be observed in the absorption spectra of certain atoms, for example calcium and magnesium.

Similarly, substituting for v in Eqn 3.13 gives the expression for the magnetic dipole moment about the field axis:

$$\boldsymbol{m} = -\frac{e\hbar}{2m_e} m_l = -\mu_B m_l. \tag{3.17}$$

Note that the dipole moment vector points in the opposite direction to the angular momentum vector, because the charge on the electron is negative. The corresponding expression for the magnitude of the *total* orbital contribution to the magnetic moment (not projected onto the field axis) is $\boldsymbol{m} = \mu_B \sqrt{l(l+1)}$.

The example of a normal Zeeman splitting of a transition between an s orbital and a p orbital is shown in Fig. 3.3. In the absence of an applied field, the s and p orbitals each have one energy level. The s energy level does not split when a field is applied, since the s electron has no orbital angular momentum and therefore no orbital magnetic moment. The p level, on the other hand, splits into three, corresponding to m_l values of $-1, 0$ and 1. As a result three lines are observed in the normal Zeeman spectrum.

3.4 Electron spin

To fully specify the state of an electron in an atom, we need to include two more quantum numbers, associated with the spin of the electron around its own axis.

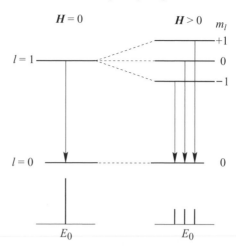

Figure 3.3 Normal Zeeman effect for a transition between s and p orbitals. The upper part of the figure shows the allowed transitions, with and without an external magnetic field. The lower part of the figure shows the corresponding absorption or emission spectra.

The spin of an electron is not predicted by the Schrödinger equation because it is the result of relativistic effects which are not included in the Schrödinger equation. If instead we had solved the relativistic Dirac equation, these additional quantum numbers would have fallen out naturally, but the mathematics would have been much more complicated!

The first new quantum number, the spin quantum number, is labeled s, and always has the value $\frac{1}{2}$. The magnitude of the spin angular momentum of an individual electron, $|S|$, is given by

$$|S| = \sqrt{s(s+1)}\hbar = \frac{\sqrt{3}}{2}\hbar. \tag{3.18}$$

This is analogous to our earlier expression for the magnitude of the orbital angular momentum, $|L|$.

The final quantum number, m_s, is the spin analog to the magnetic quantum number, m_l. It arises because the spin angular momentum with respect to a magnetic field is quantized, with m_s allowed to take values of $-\frac{1}{2}$ and $+\frac{1}{2}$ only. The component of angular momentum along the field direction is given by $m_s\hbar = \pm\hbar/2$. Again we see that the component of spin angular momentum along the applied field is smaller than the spin angular momentum magnitude. Therefore the spin angular momentum vector cannot point directly along the applied field, and instead it precesses on a cone about the field axis as shown in Fig. 3.4.

By analogy with our earlier derivation of the orbital magnetic moment along the field direction, we might expect that the spin magnetic moment be given by

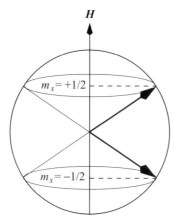

Figure 3.4 Component of angular momentum along the magnetic field direction for an s orbital (with spin quantum number $s = \frac{1}{2}$). The radius of the circle is $\frac{\sqrt{3}}{2}\hbar$.

$m = -\mu_{\mathrm{B}}m_s$. In fact this assumption is incorrect, and the theory of quantum electrodynamics instead gives

$$m = -g_{\mathrm{e}}\mu_{\mathrm{B}}m_s, \tag{3.19}$$

where $g_{\mathrm{e}} = 2.002\,319$ is called the g-factor of the electron. It will be sufficient for our purposes to take $g_{\mathrm{e}} = 2$, so that the spin magnetic moment along the field direction of a single electron is one Bohr magneton. The fact that g_{e} is not unity manifests itself in a number of important ways, and in particular it will show up later in our discussion of the anomalous Zeeman effect.

3.4.1 Pauli exclusion principle

One consequence of the well-known Pauli exclusion principle is that no two electrons can have the same values for all the quantum numbers. As a consequence, a maximum of two electrons may occupy each atomic orbital, and two electrons in the same atomic orbital (with the same values of n, l and m_l) must always have opposite spin, so that their m_s values are different.

3.5 Spin–orbit coupling

As we saw in the examples of Chapter 2, an electron with orbital angular momentum behaves like a circulating electric current, and so has a magnetic moment with an associated magnetic field. In addition, an electron possesses a magnetic moment as a result of its spin. The magnetic moment from the electron's spin interacts with the magnetic field from its orbital motion. The resulting interaction is called the spin–orbit coupling.

The magnitude of the spin–orbit coupling is determined by the charge on the atomic nucleus, which in turn depends on the atomic number, Z. This can be understood by pretending that the electron is fixed in space, with the nucleus orbiting around it, rather than the other way around. The current generated by the circulating nucleus is stronger for a larger nuclear charge. In fact the spin–orbit interaction is proportional to Z^4.[6] As a result the spin–orbit interaction is almost negligible in the hydrogen atom, but increases rapidly with atomic number.

The way in which we calculate the *total* angular momentum of all the electrons in an atom, given the l and s quantum numbers of the individual electrons, depends on the relative magnitudes of the orbit–orbit, spin–orbit and spin–spin couplings. In the remainder of this section we will discuss two different schemes for estimating the total angular momentum of a many-electron atom. This is not straightforward, but it is very important, since it's the *total* angular momentum of the electrons which determines the magnetic moment of an atom. And that, after all, is what we are interested in!

3.5.1 Russell–Saunders coupling

In light atoms, where the spin–orbit interaction is weak, the coupling between the individual orbital angular momenta and the individual spins is stronger than the spin–orbit coupling. Therefore the best way to calculate the total angular momentum is to first combine the orbital angular momenta of all the individual electrons (by vector addition) to obtain the total orbital momentum, and their spin angular momenta to obtain the total spin momentum. The total spin and orbital components are then combined to obtain the total angular momentum. The rules for determining the allowed values of the total orbital quantum number, L, given the l quantum numbers of the electrons, are rather complicated, so we won't derive them here.[†] Rather we'll give an example for an atom with just two electrons, with orbital quantum numbers l_1 and l_2 respectively. In this case the allowed values of L are given by the so-called *Clebsch–Gordan* series:

$$L = l_1 + l_2, l_1 + l_2 - 1, \ldots, |l_1 - l_2|. \tag{3.20}$$

So if we have two electrons, one with $l = 1$, and the other with $l = 2$, our allowed L values are 3, 2 and 1. By analogy with the m_l values defined for individual electrons, we define a total M_L for atoms that can range from $-L, -L + 1, \ldots,$ to $+L$, and gives the value of total orbital angular momentum projected onto a specific

[†] There is a very clear discussion in the book by Atkins.[6]

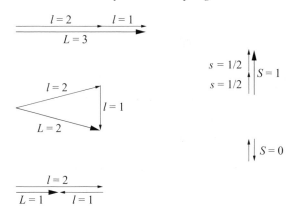

Figure 3.5 Vector summation of the orbital and spin angular momenta for two electrons with $l = 1, s = \frac{1}{2}$ and $l = 2, s = \frac{1}{2}$, to give the total atomic orbital and spin angular momentum quantum numbers, L and S.

direction. Similarly the spins are combined into a total spin quantum number

$$S = s_1 + s_2, s_1 - s_2, \tag{3.21}$$

and $M_S = -S, -S + 1, \ldots, +S$. The allowed S values for a two-electron atom are 1 or 0, with corresponding M_S values of $-1, 0, +1$ and 0 respectively.

The vector addition process is illustrated in Fig. 3.5, for the case of two electrons, one with $l = 1$, and the other with $l = 2$.

The total angular momentum quantum number, J, is then determined by vector addition of the total atomic spin and orbital angular momenta,

$$J = L + S, L + S - 1, \ldots, |L - S|. \tag{3.22}$$

For our two-electron example, the allowed values of J range from 4 to zero, and the corresponding values of M_J are $-4, -3, \ldots, 0, \ldots, 4$. The magnitude of the total atomic angular momentum, $|J|$, is then equal to $\sqrt{J(J + 1)}\hbar$, and the projection onto the magnetic field direction is $M_J\hbar$.

This scheme is known as Russell–Saunders coupling.[7] Two important points are worth noting here. First, the energy differences between states having different J values but the same L and S are small compared with those between levels having different L or S values.

Second, for a filled shell of electrons, L, S and J are equal to zero, so there is no net angular momentum and hence no contribution to the permanent magnetic dipole moment. For atoms with incomplete outer shells of electrons, we only have to consider the incomplete outer shells in calculating J, L and S. If an atom has no incomplete shells (for example in the noble gas atoms), there is no permanent dipole moment. Such atoms are called *diamagnetic*, and we will look at their properties in Chapter 4.

3.5.2 Hund's rules

The German physicist Friedrich Hund came up with a set of three rules for identifying the lowest-energy configuration for the electrons in a partially filled shell.[8] Hund's rules assume that angular momentum states are well described by Russell–Saunders coupling, therefore they do not predict the correct arrangement of electrons in the heaviest atoms.

Hund's first rule states that the electrons maximize their total spin, S. This means that the electrons will occupy orbitals with one electron per orbital, and all the spins parallel, until all the orbitals contain one electron each. The electrons are then forced to 'pair up' in orbitals, in pairs of opposite spin. This can be understood qualitatively, because electrons with the same spin are required (by the Pauli exclusion principle) to avoid each other. Therefore the repulsive Coulomb energy is less between electrons of the same spin, and the energy is lower.

The second rule states that, for a given spin arrangement, the configuration with the largest total atomic orbital angular momentum, L, lies lowest in energy. The basis for this rule is that if the electrons are orbiting in the same direction (and so have a large total angular momentum) they meet less often than if they orbit in opposite directions. Therefore their repulsion is less on average when L is large.

Finally, for atoms with less than half-full shells, the lowest-energy electronic configuration is the one with the lowest value of J (i.e. $J = |L - S|$). When the shell is more than half full the opposite rule holds – the arrangement with the highest J ($= |L + S|$) has the lowest energy. The origin of the rule is the spin–orbit coupling – and is to do with the fact that oppositely oriented dipole moments have a lower energy than those which are aligned parallel with each other.

As an example, let's look at the Mn^{2+} ion. This has five 3d electrons, and since there are five 3d orbitals, the electrons are able to occupy each orbital individually with parallel spins, as shown in Fig. 3.6. So $S = \frac{5}{2}$. As a result of maximizing the total spin, we've put an electron in each of the d orbitals, with m_l equal to $-2, -1$, 0, 1 and 2 respectively. So the sum of the m_l values is zero, and as a result L must be zero. This makes the calculation of J straightforward – if $L = 0$ then $J = S = \frac{5}{2}$.

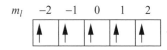

Figure 3.6 Lowest-energy arrangement of the valence electrons among the 3d orbitals for the Mn^{2+} ion.

3.5.3 jj coupling

The Russell–Saunders coupling scheme fails in heavy atoms such as the actinides because the spin and orbital angular momenta of individual electrons couple strongly. The orbital and spin angular momenta of each electron, i, combine to give the resultant total angular momenta per electron

$$j_i = l_i + s_i. \qquad (3.23)$$

The resultant j_is then interact weakly via electrostatic coupling of their electron distributions to form a resultant total angular momentum

$$J = \Sigma_i j_i. \qquad (3.24)$$

In the *jj coupling scheme*, the total orbital angular momentum quantum number, L, and the total spin angular momentum number, S, are not specified. Again, filled shells have no net angular momentum, J, and so atoms with entirely filled shells are diamagnetic.

3.5.4 The anomalous Zeeman effect

In Section 3.3 we outlined the normal Zeeman effect. In fact only atoms with a total spin angular momentum equal to zero show the normal Zeeman effect. Much more common is the so-called anomalous Zeeman effect, which gives a more complex arrangement of lines in the spectrum, and is a consequence of spin–orbit coupling. The additional complexity arises because the splittings of the upper and lower levels of the transition are unequal. The ultimate reason for the unequal splittings is the anomalous g-factor of the electron.

Because the electron g-factor, g_e, is 2 rather than 1, the total angular momentum, J, and the total magnetic moment, m, of the atom are not collinear. So the magnitude of the total atomic magnetic moment along the field axis (which determines the energy change in the presence of a magnetic field) is a function of S, L and J, rather than just of J. In fact, if we work through the mathematics[6] we obtain

$$m = -g\mu_B M_J. \qquad (3.25)$$

Here

$$g = 1 + \frac{J(J+1) + S(S+1) - L(L+1)}{2J(J+1)} \qquad (3.26)$$

is called the Landé g-factor, and $M_J = J, J-1, \ldots, -J$ is the quantum number representing the projection of the total angular momentum, J, onto the field axis. Again, the corresponding expression for the magnitude of the total magnetic moment is $m = g\mu_B \sqrt{J(J+1)}$.

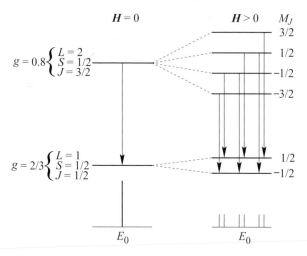

Figure 3.7 Example of a transition in the anomalous Zeeman effect.

When $S = 0$, then $g = 1$ (because $J = L$) and so the magnetic moment is independent of L and the upper and lower levels are split by the same amount. In
this case we observe the normal Zeeman effect, as we described in Section 3.3.
However, when $S \neq 0$, the value of g depends on both L and S, and so the upper
and lower levels in the spectroscopic transition are split by different amounts. This
is illustrated in Fig. 3.7. Note that the allowed transitions are further restricted by
the angular momentum conservation selection rules $\Delta M_J = 0$ (corresponding to
emission of linearly polarized light) or ± 1 (circularly polarized light).

If the applied field is very strong, the coupling between S and L can be broken
in favor of their direct coupling to the magnetic field. S and L then precess independently about the field direction. The electromagnetic field which induces the
electronic transition couples only to the orbital distribution of the electrons, and so
the presence of the spin does not show up in the transitions. Therefore the spectrum
switches back from the anomalous to the normal Zeeman effect. This change in the
spectrum is known as the Paschen–Back effect.[9]

Homework

Exercises

3.1   Calculate the allowed values of the magnetic moment along the field axis of an atom
which has $J = 1$ and $g = 2$.

3.2a   What is the electronic configuration of an Fe^{2+} ion? (Note that the transition metals
give up their 4s electrons before their 3d electrons on ionization.)

3.2b   Use Hund's rules to determine the values of S, L and J in the ground state of an
Fe^{2+} ion.

3.2c Calculate the Landé g-factor using Eqn 3.26.

3.2d Calculate the total magnetic moment of an Fe^{2+} ion, $g\sqrt{J(J+1)}\mu_B$, and the magnetic moment along the field direction, $gM_J\mu_B$. Compare your result with your answer to Exercise 1.3b.

3.2e Try calculating the total magnetic moment using the value of S determined in 3.2b, but assuming that $L = 0$ (so $J = S$). In fact the measured value is $5.4\mu_B$. More about this later!

Further reading

P.W. Atkins *Molecular quantum mechanics*. Oxford University Press, 1999.

4

Diamagnetism

"A sensitive compass having a Bi needle would be ideal for the young man going west or east, for it always aligns itself at right angles to the magnetic field."

William H. Hayt Jr., *Engineering electromagnetics.*
McGraw-Hill, 1958.

In the previous chapter we studied two contributions to the magnetic moment of atoms – the electron spin and orbital angular momenta. Next we are going to investigate the third (and final) contribution to the magnetic moment of a free atom. This is the *change* in orbital motion of the electrons when an external magnetic field is applied.

The change in orbital motion due to an applied field is known as the diamagnetic effect, and it occurs in *all* atoms, even those in which all the electron shells are filled. In fact diamagnetism is such a weak phenomenon that *only* those atoms which have no net magnetic moment as a result of their shells being filled are classified as diamagnetic. In other materials the diamagnetism is overshadowed by much stronger interactions such as ferromagnetism or paramagnetism.

4.1 Observing the diamagnetic effect

The diamagnetic effect can be observed by suspending a container of diamagnetic material, such as bismuth, in a magnetic field gradient, as shown in Fig. 4.1. The energy of a diamagnetic material is *increased* by the presence of a field, and so the cylinder swings away from the high field region, towards the region of lower field (the north pole in the figure). Although Bi is one of the strongest diamagnetic materials, the deflection is quite small because the diamagnetic effect is always weak.

Although the diamagnetic effect might seem counter-intuitive, it actually makes perfect sense! When the magnetic field is turned on, extra currents are generated in the atom by electromagnetic induction. Lenz's law tells us that the currents are

34

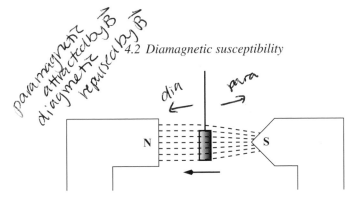

Figure 4.1 Experimental setup to observe diamagnetism.

induced in the direction which opposes the applied field, so the induced magnetic moments are directed opposite to the applied field. So, the stronger the field, the more 'negative' the magnetization. And, even if the magnetic moments of the free atom cancel out, the *changes* in magnetic moment always act to oppose the field, which explains why atoms with no net magnetic moment still show a diamagnetic effect.

4.2 Diamagnetic susceptibility

In Chapter 2 we introduced the concept of susceptibility – the variation in magnetization of a material with applied magnetic field. We stated that the susceptibility of a diamagnetic material is negative, that is the magnetization decreases as the magnetic field is increased.

Next let's compute an expression for the value of the diamagnetic susceptibility, χ, in a free atom. We'll use a classical derivation (in fact the quantum mechanical derivation gives the same result) known as the Langevin theory,[10] which explains the negative susceptibility in terms of the motion of electrons as we discussed above.[†] The derivation is most elegant in SI units – we'll give the equivalent expression in cgs units at the end of this section.

Consider an electron orbiting perpendicular to an applied field, and generating a current in the opposite direction to its motion as shown below:

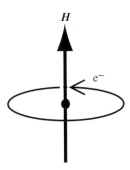

[†] In fact it is encouraging to us mere mortals that Langevin made an error in his mathematics which was later corrected by Pauli.[11]

When the magnetic field is turned on slowly from zero, the change in flux, Φ, through the current loop induces an electromotive force, ε, which acts to oppose the change in flux. The electromotive force is defined as the line integral of the electric field E around any closed path, and Faraday's law tells us that it is equal to the rate of change of magnetic flux through the path. If we take the electron orbit of radius r as our path, then

$$\varepsilon = E \times 2\pi r = -\frac{d\Phi}{dt}. \tag{4.1}$$

In fact the change in flux is achieved by decreasing the electron velocity (and as a consequence decreasing the circulating current, I). As a side effect it also decreases the magnetic moment of the loop, and it's this decrease in the magnetic moment that we observe as the diamagnetic effect. Although the electromotive force only acts while the field is *changing*, the new value of the current persists because there is no resistance to the motion of the electron. So the magnetic moment is decreased as long as the field is acting.

The torque exerted on the electron by the induced electric field is $-eEr$, and this has to equal the rate of change of angular momentum, dL/dt. So

$$\frac{dL}{dt} = -eEr = +\frac{e}{2\pi}\frac{d\Phi}{dt} = \frac{er^2\mu_0}{2}\frac{dH}{dt} \tag{4.2}$$

(because in general $\Phi = \mu H A$, where $A = \pi r^2$ is the area of the current loop, and here we take the permeability $\mu = \mu_0$ because we are considering a free atom). Integrating with respect to time from zero field, we find the change in angular momentum from turning on the field is

$$\Delta L = \frac{er^2\mu_0}{2}H. \tag{4.3}$$

This additional angular momentum makes an extra magnetic moment, which is just $-e/2m_e$ multiplied by the angular momentum. (Remember $L = m_e v a$ and $I = ev/2\pi a$, so $\boldsymbol{m} = IA = -e/2m_e L$.) So the change in magnetic moment,

$$\Delta\boldsymbol{m} = -\frac{e}{2m_e}\Delta L \tag{4.4}$$

$$= -\frac{e^2 r^2 \mu_0}{4m_e}H. \tag{4.5}$$

We see that the induced magnetic moment is proportional to the applied magnetic field, and in the opposite direction to it.

In this derivation we have assumed that the field $\boldsymbol{H}$ is perpendicular to the electron orbit. In fact in the classical description all orientations are allowed, and instead of using the orbital radius r^2, we should use use the average value of the square of the projection of r onto the field direction. This reduces the effective magnetic moment by a factor of $\frac{2}{3}$. In addition, if electrons from different atomic orbitals contribute to the diamagnetism, then we need to take the *average* value of all occupied orbital radii, $\langle r^2 \rangle_{av}$, and multiply by the number of electrons, Z. So

$$\Delta m = -\frac{Ze^2 \langle r^2 \rangle_{av} \mu_0}{6m_e} H. \tag{4.6}$$

Finally, to convert to a bulk magnetization we multiply by the number of atoms per unit volume, N. (Note that N is equal to $N_A \rho / A$, where N_A is Avagadro's number (the number of atoms per mole), ρ is the density and A is the atomic weight.) Then the diamagnetic susceptibility is given by

$$\chi = \frac{M}{H} \tag{4.7}$$

$$= -\frac{N \mu_0 Z e^2}{6m_e} \langle r^2 \rangle_{av}. \tag{4.8}$$

Note that this is dimensionless. We see that the diamagnetic susceptibility is always negative, and that there is no explicit temperature dependence. However the amount of magnetization is proportional to $\langle r^2 \rangle_{av}$, which is weakly temperature dependent. The magnitude of the diamagnetic susceptibility is around 10^{-6} per unit volume, which is very small. If we had worked in cgs units we would have ended up with almost the same expression:

$$\chi = -\frac{N Z e^2}{6m_e c^2} \langle r^2 \rangle_{av} \tag{4.9}$$

in units of emu/cm^3 Oe.

4.3 Diamagnetic substances

All the noble gases are diamagnetic, because they have filled electron shells. Also many diatomic gases are diamagnetic, because the electrons pair up in the molecular orbitals to leave no net magnetic moment. This is illustrated in Fig. 4.2 for the hydrogen (H_2) molecule. (We will discuss paramagnetic diatomic gases, such as O_2, in Chapter 5.)

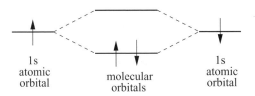

Figure 4.2 Formation of H_2 molecular orbitals from hydrogen atomic orbitals. Each hydrogen atom 1s orbital contains one electron. In the H_2 molecule, two electrons fill the lowest molecular orbital leaving no net angular momentum.

4.4 Uses of diamagnetic materials

Diamagnetic materials do not have a permanent magnetic moment and therefore do not find the wide range of applications that other magnetic materials do. However one rather interesting use arises for alloys of diamagnetic and paramagnetic materials. Paramagnetic materials have a positive susceptibility, therefore alloys containing a mixture of dia- and paramagnetic materials always have a particular composition (at each temperature) at which the magnetism exactly cancels out and the susceptibility is zero. At this composition, the alloy is completely unaffected by magnetic fields, and therefore it is used in equipment which is designed to make delicate magnetic measurements.

A novel application of diamagnetism which has been explored recently is the magnetic-field-induced alignment of liquid crystals.[12, 13] A strong magnetic field induces alignment of liquid crystals in which the diamagnetic susceptibility is *anisotropic*. Since diamagnetic materials tend to exclude magnetic flux, the liquid crystals orient themselves such that the axis with the most negative diamagnetic susceptibility is perpendicular to the field. The amount of macroscopic alignment can then be controlled by adjusting the composition of the liquid crystal to change its diamagnetic susceptibility.[12] This effect can in turn be exploited to align meso-porous inorganic materials such as silica by filling the anisotropic pores with liquid crystal surfactants.[13]

4.5 Superconductivity

The most well-known materials that show diamagnetic behavior are the supercon-ductors. These are materials which undergo a transition from a state of normal elec-trical resistivity to one of zero resistivity when cooled below a critical temperature, T_c. Below T_c, superconductors are in fact 'perfect' diamagnets, with a susceptibility of -1. They are fundamentally different from conventional diamagnets, however, in that the susceptibility is caused by *macroscopic* currents circulating in the material to oppose the applied field, rather than by changes in the orbital motion of closely bound electrons.

The science of superconductivity is extremely rich, and the details are beyond the scope of this book. However in the remainder of this chapter we will give a brief overview of some of the fundamentals.

4.5.1 The Meissner effect

If a metal such as lead, which is normally diamagnetic, is cooled in a magnetic field, then at some critical temperature, T_c, it will spontaneously exclude all magnetic flux from its interior, as illustrated in Fig. 4.3. If $\boldsymbol{B} = \mu_0(\boldsymbol{H} + \boldsymbol{M}) = 0$, then $\boldsymbol{M} = -\boldsymbol{H}$,

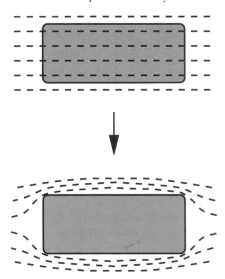

Figure 4.3 Schematic of the Meissner effect. The top diagram shows the lines of flux penetrating the material in its normal state. In the lower diagram the material has been cooled below its superconducting transition temperature, and the magnetic field has been simultaneously excluded.

and $\chi = M/H = -1$ (in SI units). And the permeability, $\mu = 1 + \chi = 0$, so the material is *impermeable* to the magnetic field. T_c is also the temperature at which the material undergoes the transition to the superconducting state.

The exclusion of flux is called the Meissner effect,[14] and is the reason that superconductors are perfect diamagnets. The circulating currents which (by Lenz's law) oppose the applied magnetic field are able to exactly cancel the applied field because the resistivity is zero in the superconducting state. This is the reason that the exclusion of flux coincides with the onset of superconductivity. The highest known critical temperatures are around 130 K, in layered copper oxide materials, (the so-called high-T_c superconductors).

4.5.2 Critical field

Even below T_c, the superconducting state can be destroyed if a high enough field is applied. The field which destroys the superconducting state at a particular temperature is called the critical field, H_c. At lower temperatures, the critical field is higher, and by definition it is zero at T_c because the superconducting state is destroyed spontaneously.

If the superconductor is carrying a current, then the field produced by the circulating charge also contributes to H_c. Therefore there is a maximum allowable current before superconductivity is destroyed. The critical current depends on the radius of the conductor and is a crucial factor in determining the technological utility of a particular superconducting material.

4.5.3 Classification of superconductors

Superconductors can be classified as type I or type II. In type I superconductors, the induced magnetization is proportional to the applied field, and a plot of M versus H has a slope of -1 all the way up to the critical field, H_c. They are always perfect diamagnets in their superconducting state. Usually type I superconductors are pure materials which tend to have low critical fields and are therefore not useful for many applications.

Type II superconductors undergo a transition from type I superconductivity to a vortex state, in which the superconductor is threaded by flux lines, at a critical field, H_{c1}. Between the lower critical field, H_{c1}, and a second, higher critical field, H_{c2}, the vortex state persists. At H_{c2} the superconducting state is destroyed and normal conductivity resumed. The advantage of type II superconductors is that H_{c2} is high enough to allow practical applications.

Schematic magnetization curves for type I and II superconductors are plotted in Fig. 4.4.

4.5.4 Applications for superconductors

SQUIDs

Superconducting quantum interference devices (or SQUIDs) are devices which are capable of measuring very small changes in magnetic field. They make use of the Josephson effect[15] in which two pieces of superconducting material are

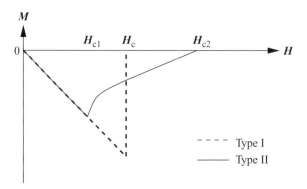

Figure 4.4 Magnetization versus applied magnetic field for type I and type II super-conductors.

separated by a very thin insulating layer. Superconducting electrons can tunnel through the barrier, but the critical current density is changed by the presence of a very small field. The SQUID uses this change in current to detect the small magnetic field.

Superconducting magnets

Materials with high critical fields, such as niobium–tin, Nb_3Sn, can sustain high current densities, and therefore generate high magnetic fields when wound into a superconducting solenoid. These superconducting magnets are used in research laboratories, and also have practical applications such as magnetic resonance imaging (MRI).

Homework

Exercises

4.1 The value of $\sqrt{\langle r^2 \rangle_{av}}$ for carbon is known from x-ray diffraction measurements to be around 0.7 Å. The density is 2220 kg/m^3. Calculate the value of susceptibility (give your answer in SI and cgs units). The measured value is -13.82×10^{-6} per unit volume. The agreement for carbon is better than that for most diamagnets. Comment on possible sources of error in the derivation.

Further reading

D. Jiles, *Introduction to magnetism and magnetic materials*. Chapman & Hall, 1996, Chapter 15.
J.R. Schrieffer, *Theory of superconductivity*. Perseus Press, 1988.
M. Tinkham, *Introduction to superconductivity*. McGraw-Hill, 1995.
P.G. De Gennes *Superconductivity of metals and alloys*. Perseus Press, 1994.

5

Paramagnetism

"A grocer is attracted to his business by a magnetic force as great as the repulsion which renders it odious to artists."

Honoré De Balzac, *Les célibataires.* 1841.

In the previous chapter we discussed the diamagnetic effect, which is observed in all materials, even those in which the constituent atoms or molecules have no permanent magnetic moment. Next we are going to discuss the phenomenon of paramagnetism, which occurs in materials that have net magnetic moments. In paramagnetic materials these magnetic moments are only weakly coupled to each other, and so thermal energy causes random alignment of the magnetic moments, as shown in Fig. 5.1(a). When a magnetic field is applied, the moments start to align, but only a small fraction is deflected into the field direction for all practical field strengths. This is illustrated in Fig. 5.1(b).

Many salts of transition elements are paramagnetic. In transition metal salts, each transition metal cation has a magnetic moment resulting from its partially filled d shell, and the anions ensure spatial separation between cations. Therefore the interactions between the magnetic moments on neighboring cations are weak. The rare earth salts also tend to be paramagnetic. In this case the magnetic moment is caused by highly localized f electrons, which do not overlap with f electrons on adjacent ions. There are also some paramagnetic metals, such as aluminum, and some paramagnetic gases, such as oxygen, O_2. All ferromagnetic materials (which we will discuss in the next chapter) become paramagnetic above their Curie temperature, when the thermal energy is high enough to overcome the cooperative ordering of the magnetic moments.

At low fields, the flux density within a paramagnetic material is directly proportional to the applied field, so the susceptibility, $\chi = M/H$, is approximately constant. Generally χ is between around 10^{-3} and 10^{-5}. Because the susceptibility is only slightly greater than zero, the permeability is slightly greater than 1 (unlike

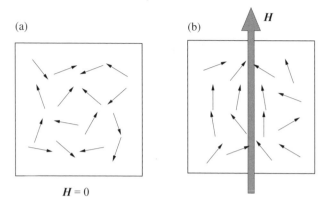

Figure 5.1 Schematic of the alignment of magnetic moments in a paramagnetic material: (a) shows the alignment in the absence of an external field, and (b) shows the response when a field of moderate strength is applied.

diamagnets, where it was slightly less than 1). In many cases, the susceptibility is inversely proportional to the temperature. This temperature dependence of χ can be explained by the Langevin localized moment model[10] which we will discuss in the next section. However in some metallic paramagnets the susceptibility is independent of the temperature – these are the Pauli paramagnets. The paramagnetism in the Pauli paramagnets results from a quite different mechanism, and is well described by the band structure theory of collective electrons. We will discuss Pauli paramagnetism in Section 5.4.

5.1 Langevin theory of paramagnetism

The Langevin theory explains the temperature dependence of the susceptibility in paramagnetic materials by assuming that the non-interacting magnetic moments on atomic sites are randomly oriented as a result of their thermal energy. When an external magnetic field is applied, the orientation of the atomic moments shifts slightly towards the field direction as shown schematically in Fig. 5.1. We will derive the expression for the susceptibility using a classical argument, then extend it to the quantum mechanical case at the end of the derivation.

For a moment which makes an angle θ to the applied field H, the probability of occupying an energy state, E, is (by Boltzmann statistics)

$$e^{-E/k_{B}T} = e^{m \cdot H/k_{B}T} = e^{mH \cos\theta/k_{B}T}. \tag{5.1}$$

(Here the un-bold m and H represent the magnitude of the magnetic moment and field vectors respectively, and k_{B} is Boltzmann's constant). We can calculate the number of moments lying between angles θ and $\theta + d\theta$ with respect to the field

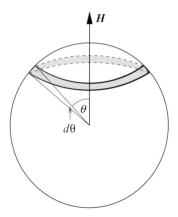

Figure 5.2 The fraction of paramagnetic moments between angles θ and $\theta + d\theta$ around an axis is equal to the fractional area that the angle $d\theta$ sweeps out on the surface of the sphere as shown.

H, by noticing that it is proportional to the fractional surface area of a surrounding sphere, as shown in Fig. 5.2. This fractional surface area, $dA = 2\pi r^2 \sin\theta d\theta$.

So the overall probability, $p(\theta)$, of an atomic moment making an angle between θ and $\theta + d\theta$ is

$$p(\theta) = \frac{e^{mH\cos\theta/k_BT}\sin\theta d\theta}{\int_0^\pi e^{mH\cos\theta/k_BT}\sin\theta d\theta}, \tag{5.2}$$

where the denominator is the *total* number of atomic magnetic moments, and the factors of $2\pi r^2$ cancel out.

Each moment contributes an amount $m\cos\theta$ to the magnetization parallel to the magnetic field, and so the magnetization from the whole system is

$$M = Nm\langle\cos\theta\rangle \tag{5.3}$$

$$= Nm \int_0^\pi \cos\theta p(\theta)\, d\theta \tag{5.4}$$

$$= Nm\frac{\int_0^\pi e^{mH\cos\theta/k_BT}\cos\theta \sin\theta d\theta}{\int_0^\pi e^{mH\cos\theta/k_BT}\sin\theta d\theta}. \tag{5.5}$$

$$\tag{5.6}$$

Carrying out the nasty integrals (or looking them up in tables!) gives

$$M = Nm\left[\coth\left(\frac{mH}{k_BT}\right) - \frac{k_BT}{mH}\right] \tag{5.7}$$

$$= NmL(\alpha), \tag{5.8}$$

where $\alpha = mH/k_BT$ and $L(\alpha) = \coth(\alpha) - 1/\alpha$ is called the Langevin function. The form of $L(\alpha)$ is shown in Fig. 5.3. If α were made large enough, for example

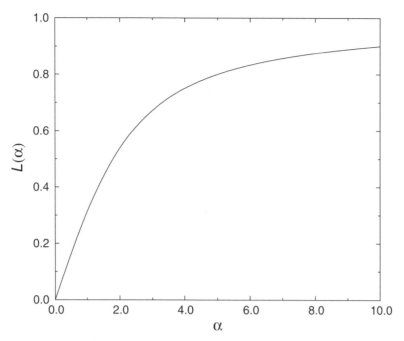

Figure 5.3 The Langevin function, $L(\alpha)$.

by applying a very large field, or lowering the temperature towards zero kelvin, then M would approach Nm, and complete alignment of the magnetic spins could be achieved.

Now what about our earlier statement that $\chi \propto 1/T$? We were expecting to see $M =$ some constant $\times H/T$ and we've ended up with something far more complicated. Well, the Langevin function can be expanded as a Taylor series:

$$L(\alpha) = \frac{\alpha}{3} - \frac{\alpha^3}{45} + \cdots . \tag{5.9}$$

So, keeping only the first term, (which dominates at all practical fields and temperatures since α is very small),

$$M = \frac{Nm\alpha}{3} = \frac{Nm^2}{3k_B} \frac{H}{T}. \tag{5.10}$$

(The equivalent expression in SI units is $M = (N\mu_0 m^2/3k_B)(H/T)$, since $E = -\mu_0 m \cdot H$). The magnetization is proportional to the applied field and inversely proportional to the temperature, as we expected. This gives the susceptibility

$$\chi = \frac{M}{H} = \frac{Nm^2}{3k_B T} = \frac{C}{T}, \tag{5.11}$$

where $C = Nm^2/3k_B$ is a constant. This is Curie's law – the susceptibility of a paramagnet is inversely proportional to the temperature.

So far we have assumed that the magnetic dipole moment can take all possible orientations with respect to the applied magnetic field, whereas in reality it can have only discrete orientations because of spatial quantization. If we incorporate the quantization into the derivation of the total magnetization, we obtain:

$$M = NgJ\mu_B \left[\frac{2J+1}{2J} \coth \left(\frac{2J+1}{2J}\alpha \right) - \frac{1}{2J} \coth \left(\frac{\alpha}{2J} \right) \right] \qquad (5.12)$$

$$= NgJ\mu_B B_J(\alpha). \qquad (5.13)$$

$B_J(\alpha)$ is the Brillouin function, which is equal to the Langevin function in the limit that $J \to \infty$. The Brillouin function can also be expanded in a Taylor series:

$$B_J(\alpha) = \frac{J+1}{3J}\alpha - \frac{[(J+1)^2 + J^2](J+1)}{90J^3}\alpha^3 + \cdots . \qquad (5.14)$$

Here $\alpha = Jg\mu_B H/k_B T$.

Keeping only the first term in the expansion, the quantum mechanical expression for the susceptibility becomes

$$\chi = \frac{Ng^2 J(J+1)\mu_B^2}{3k_B T} = \frac{C}{T}. \qquad (5.15)$$

Again, to obtain the susceptibility in SI units, this expression is multiplied by μ_0. The overall form of the response is the same as in the classical, case, but this time the proportionality constant, C, is given by $Ng^2 J(J+1)\mu_B^2/3k_B = Nm_{\text{eff}}^2/3k_B$ where $m_{\text{eff}} = g\sqrt{J(J+1)}\mu_B$.

5.2 The Curie–Weiss law

In fact many paramagnetic materials do not obey the Curie law which we just derived, but instead follow a more general temperature dependence given by the Curie–Weiss law:

$$\chi = \frac{C}{T - \theta}. \qquad (5.16)$$

Paramagnets which follow the Curie–Weiss law undergo spontaneous ordering and become ferromagnetic below some critical temperature, the Curie temperature, T_C (which we'll see later is for all practical purposes equal to θ).

In our derivation of the Curie law we assumed that the localized atomic magnetic moments do not interact with each other at all – they are just reoriented by the applied magnetic field. Weiss explained the observed Curie–Weiss behavior by postulating the existence of an internal interaction between the localized moments which he called a 'molecular field'. He did not speculate as to the origin of his molecular field, beyond suggesting that it is a mutual interaction between the electrons which tends to align the dipole moments parallel to each other. (We can't really criticize Weiss for this – remember that the electron had been discovered only 10 years earlier, and quantum mechanics hadn't been 'invented' yet!)

Weiss assumed that the intensity of the molecular field is directly proportional to the magnetization:

$$\boldsymbol{H}_{\mathrm{W}} = \gamma \boldsymbol{M}, \tag{5.17}$$

where γ is called the *molecular field constant*. So the total field acting on the material is

$$\boldsymbol{H}_{\mathrm{tot}} = \boldsymbol{H} + \boldsymbol{H}_{\mathrm{W}}. \tag{5.18}$$

We just derived

$$\chi = \frac{M}{H} = \frac{C}{T}, \tag{5.19}$$

so, replacing $\boldsymbol{H}$ by $\boldsymbol{H}_{\mathrm{tot}} = \boldsymbol{H} + \gamma \boldsymbol{M}$,

$$\frac{M}{H + \gamma M} = \frac{C}{T} \tag{5.20}$$

or

$$M = \frac{CH}{T - C\gamma}. \tag{5.21}$$

Therefore

$$\chi = \frac{M}{H} = \frac{C}{T - \theta}, \tag{5.22}$$

the Curie–Weiss law!

When $T = \theta$ there is a divergence in the susceptibility, which corresponds to the phase transition to the spontaneously ordered phase. A positive value of θ indicates that the molecular field is acting in the same direction as the applied field, and tending to make the elementary magnetic moments align parallel to one another and to the applied field. This is the case in a ferromagnetic material.

We can estimate the size of the Weiss molecular field. Below the critical temperature, T_{C}, paramagnetic materials exhibit ferromagnetic behavior. Above T_{C}, the thermal energy outweighs $\boldsymbol{H}_{\mathrm{W}}$, and the ferromagnetic ordering is destroyed. Therefore at T_{C}, the interaction energy, $\mu_{\mathrm{B}}\boldsymbol{H}_{\mathrm{W}}$, must be approximately equal to the thermal energy, $k_{\mathrm{B}}T_{\mathrm{C}}$. So $\boldsymbol{H}_{\mathrm{W}} \approx k_{\mathrm{B}}T_{\mathrm{C}}/\mu_{\mathrm{B}} \approx 10^{-16}10^3/10^{-20} \approx 10^7$ Oe. This is extremely large! In the next chapter we will apply Weiss's molecular field theory *below* the Curie temperature to understand the ferromagnetic phase, and we will discuss the origin of the molecular field.

The Langevin theory and the Curie–Weiss law give accurate descriptions of many paramagnetic materials. Next, we will look at two cases where they don't do so well. The first is not really a problem with the theory, but a difference in the size of the measured and predicted magnetic moments of the ions. The second is an example of a class of materials (the Pauli paramagnets) where the assumptions of the Langevin localized moment theory no longer apply.

5.3 Quenching of orbital angular momentum

The total magnetization in a paramagnet depends on the magnetic moment, m, of the constituent ions. Once we know the g-factor of an ion, and the J value, we can calculate it's magnetic moment – it's just $m = g\mu_B\sqrt{J(J+1)}$. (This after all was the whole purpose of Chapter 3!) In general this formula works very well for paramagnetic salts, even though the ions have formed into crystals and are no longer 'free'. As an example we show the calculated and experimental values for the rare earth ions in Table 5.1. In all cases (except for the Eu^{3+} ion) the agreement is very good. In Eu^{3+} the calculated magnetic moment for the ground state is zero, however there are low-lying excited states which do have a magnetic moment and which are partially occupied at practical temperatures. Averaging over

Table 5.1 *Calculated and measured effective magnetic moments for the rare earth ions.*

(From Ref. 16, Kittel, *Introduction to solid state physics*, 7th edn. Copyright 1995 John Wiley & Sons, Inc. Reprinted by permission of John Wiley & Sons, Inc.)

ion	configuration	$g\sqrt{J(J+1)}$	m/μ_B
Ce^{3+}	$4f^1 5s^2 5p^6$	2.54	2.4
Pr^{3+}	$4f^2 5s^2 5p^6$	3.58	3.5
Nd^{3+}	$4f^3 5s^2 5p^6$	3.62	3.5
Pm^{3+}	$4f^4 5s^2 5p^6$	2.68	–
Sm^{3+}	$4f^5 5s^2 5p^6$	0.84	1.5
Eu^{3+}	$4f^6 5s^2 5p^6$	0.00	3.4
Gd^{3+}	$4f^7 5s^2 5p^6$	7.94	8.0
Tb^{3+}	$4f^8 5s^2 5p^6$	9.72	9.5
Dy^{3+}	$4f^9 5s^2 5p^6$	10.63	10.6
Ho^{3+}	$4f^{10} 5s^2 5p^6$	10.60	10.4
Er^{3+}	$4f^{11} 5s^2 5p^6$	9.59	9.5
Tm^{3+}	$4f^{12} 5s^2 5p^6$	7.57	7.3
Yb^{3+}	$4f^{13} 5s^2 5p^6$	4.54	4.5

Table 5.2 *Calculated and measured effective magnetic moments for the first row transition metal ions.*

(From Ref. 16, Kittel, *Introduction to solid state physics*, 7th edn. Copyright 1995 John Wiley & Sons, Inc. Reprinted by permission of John Wiley & Sons, Inc.)

ion	configuration	$g\sqrt{J(J+1)}$	$g\sqrt{S(S+1)}$	m/μ_B
Ti^{3+}, V^{4+}	$3d^1$	1.55	1.73	1.8
V^{3+}	$3d^2$	1.63	2.83	2.8
Cr^{3+}, V^{2+}	$3d^3$	0.77	3.87	3.8
Mn^{3+}, Cr^{2+}	$3d^4$	0.00	4.90	4.9
Fe^{3+}, Mn^{2+}	$3d^5$	5.92	5.92	5.9
Fe^{2+}	$3d^6$	6.70	4.90	5.4
Co^{2+}	$3d^7$	6.63	3.87	4.8
Ni^{2+}	$3d^8$	5.59	2.83	3.2
Cu^{2+}	$3d^9$	3.55	1.73	1.9

the calculated magnetic moments for these excited states gives a value which is in agreement with the measured value.

However for the first row transition metals, things do not work out quite so nicely, and in fact the measured magnetic moment is closer to that which we would calculate if we completely ignored the *orbital* angular momentum of the electrons. Table 5.2 lists the measured magnetic moments, and the calculated values using the total and spin-only angular momenta. It's clear that the spin-only values are in much better agreement with experiment than the values calculated using the total angular momentum. This phenomenon is known as *quenching* of the orbital angular momentum, and is a result of the electric field generated by the surrounding ions in the solid. Qualitatively, these electric fields cause the orbitals to be coupled strongly to the crystal lattice, so that they are not able to reorient towards an applied field, and so do not contribute to the observed magnetic moment. The spins, on the other hand are only weakly coupled to the lattice – the result is that only the spins contribute to the magnetization process, and, consequently, to the resultant magnetic moment of the specimen. For a more detailed discussion see Ref. 16.

5.4 Pauli paramagnetism

In the Langevin theory we assumed that the electrons in the partially occupied valence shells (which cause the net atomic magnetic moments) were fully localized on their respective atoms. We know that, in metals, the electrons are able to wander through the lattice and give rise to electrical conductivity. So the localized moment approximation is unlikely to be a good one. This is in fact the case, and in paramagnetic metals we do not see the $1/T$ susceptibility dependence characteristic of Langevin paramagnets. Instead the susceptibility is more or less independent of temperature – a phenomenon known as Pauli paramagnetism. Before we can explain Pauli paramagnetism we need to understand the concept of energy bands.

5.4.1 Energy bands in solids

We saw in Chapter 3 that the electrons in atoms occupy discrete energy levels known as atomic orbitals. When atoms are brought together to form a solid, the wavefunctions of their outermost valence electrons overlap and the electronic configuration is altered. In fact, each discrete orbital energy of the free atom contributes to a continuous *band* of allowed energy levels in the solid. The greater the amount of overlap between the wavefunctions, the broader the band. So the valence electrons occupy rather broad bands, whereas the bands produced from the more tightly bound core electrons are narrow.

The band formation process is illustrated for sodium in Fig. 5.4. The atomic orbital energy levels, corresponding to infinitely separated Na atoms, are shown on the left of the figure. A free Na atom has fully occupied 1s, 2s and 2p sub-shells, and a single electron in the 3s orbital. The 3p orbital is empty in the ground state. When the atoms are brought together the wavefunctions of the valence electrons start to overlap and band formation occurs. At the equilibrium bonding distance,

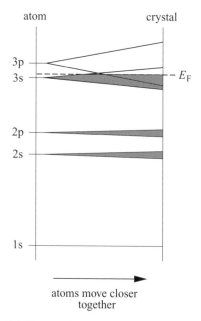

Figure 5.4 Energy band formation in sodium metal.

the bands derived from the 3s and 3p atomic orbitals are so wide that they overlap. The lower-lying core electrons have very little wavefunction overlap, and correspondingly narrow bands.

Just as in free atoms, the electrons in solids occupy the energy bands starting with those of the lowest energy and working up. The bands which derived from filled atomic orbitals are filled completely. In sodium, the electrons which occupied the 3s orbital in the atom now occupy the overlapping 3s–3p bands – a fraction are in 3s states and the remainder are in 3p states. (We'll see in the next chapter that this overlapping of energy bands has an important effect in determining the average atomic magnetic moments in ferromagnetic transition metals.)

The highest energy level which is filled with electrons at zero kelvin is called the Fermi energy, E_F. One characteristic of paramagnetic metals is that the energy states for up- and down-spin electrons are the same, and so the energy levels at the Fermi energy are identical for up- and down-spins. (We'll see later that this is not the case in ferromagnetic metals, where there are more electrons of one spin, giving rise to a net magnetic moment.) This is illustrated schematically in Fig. 5.5(a). (Remember that the energy levels really form a continuous band; we have drawn discrete levels for clarity). When a magnetic field is applied, however, those electrons with their magnetic moments aligned parallel to the field have a lower energy than those which are antiparallel. (If the field is applied in the up direction, then the down-spin electrons have lower energy than the up-spin electrons, since the negative electronic charge makes the magnetic moment point in the opposite direction to the spin.) So there is a tendency for the antiparallel electrons to try and reorient themselves parallel to the field. However, because of the Pauli exclusion principle, the only way that they can do this is by moving into one of the vacant parallel-moment states, and only those electrons close to the Fermi level have sufficient energy to

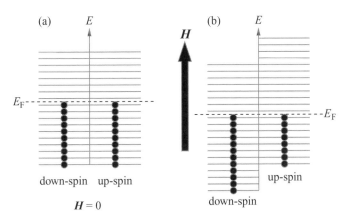

Figure 5.5 Adjustment of electronic energies in a paramagnetic metal when a magnetic field is applied.

do this. For the lower-lying electrons, the energy gained by realignment would be outweighed by that required to promote the electron to the vacant state. This is illustrated in Fig. 5.5(b). Figure 5.5 also shows that Pauli paramagnets develop an overall magnetization when a magnetic field is applied.

Before we can quantify this change in induced magnetization, and derive an expression for the susceptibility, we need to have a model for the electrons in a metal. In the next section we'll derive the so-called 'free electron theory' which describes the properties of many simple metals well.

5.4.2 Free electron theory of metals

The free electron theory assumes that the valence electrons in a solid are completely ionized from their parent atoms, and behave like a 'sea' of electrons wandering around in the solid. These electrons, the free electron gas, move in the average field created by all the other electrons and the ion cores, and, for each electron, the repulsive potential from the other electrons is assumed to exactly cancel out the attractive ion core potentials. Despite this huge approximation, the free electron theory yields surprisingly good results for simple metals. (The reasons for the success of the free electron model are rather subtle and confused condensed matter physicists for a long time. Unfortunately we don't have time to go into them here – there is an excellent discussion in the review by Cohen.[17])

The Schrödinger equation for free electrons includes only a kinetic energy term, because by definition the potential energy is zero. So, in three dimensions it is

$$-\frac{\hbar^2}{2m_e}\left(\frac{\partial^2}{\partial x^2} + \frac{\partial^2}{\partial y^2} + \frac{\partial^2}{\partial z^2}\right)\psi_k(\mathbf{r}) = E_k\psi_k(\mathbf{r}). \tag{5.23}$$

The most straightforward method for solving this equation is to pretend that the electrons are confined to a cube of edge length L, and that they satisfy periodic boundary conditions. Then the solutions are traveling plane waves,

$$\psi_k(\mathbf{r}) = e^{i\mathbf{k}\cdot\mathbf{r}} \tag{5.24}$$

provided that the wavevector $\mathbf{k}$ satisfies

$$k_x, k_y, k_z = \pm\frac{2n\pi}{L} \tag{5.25}$$

where n is any positive integer. In a macroscopic solid, L is very large and so the spectrum of allowed $\mathbf{k}$ values is effectively continuous.

Substituting $\psi_k(\mathbf{r})$ back into the Schrödinger equation gives us the energy eigenvalues

$$E_k = \frac{\hbar^2}{2m_e}\left(k_x^2 + k_y^2 + k_x^2\right). \tag{5.26}$$

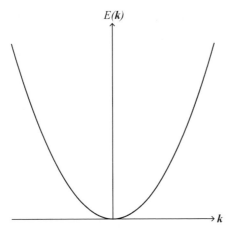

Figure 5.6 Energy versus wavevector for a free electron gas.

The energy is quadratic in the wavevector, as shown in Fig. 5.6.

Now as we saw in Section 5.4.1, the important quantity for determining the response of a Pauli paramagnet to a magnetic field is the *number* of electrons close to the Fermi energy level which are able to reverse their spin when a field is applied. So next let's derive an expression for the *density of states* (that is the number of electron energy levels per unit energy range) at the Fermi level.

We just showed that the energy of a particular k-state is given by $E = (\hbar^2/2m_e)k^2$. In particular the Fermi energy is given by $E = (\hbar^2/2m_e)k_F^2$, where k_F is the wavevector of the highest filled state, and lies on a sphere of volume $\frac{4}{3}\pi k_F^3$, within which all states are filled. We also know that the components of the k-vector, k_x, k_y and k_z, are quantized in multiples of $2\pi/L$. So the volume occupied by a single quantum state in k-space must be $(2\pi/L)^3$. Therefore the total number of electrons, which is equal to twice the number of occupied orbitals (one electron each of up- and down-spin), is given by

$$N = \frac{\text{volume of Fermi sphere}}{\text{volume per } k\text{-state}} \times 2 \tag{5.27}$$

$$= \left(\frac{\frac{4}{3}\pi k_F^3}{\left(\frac{2\pi}{L} \right)^3} \right) \times 2 \tag{5.28}$$

$$= \frac{V}{3\pi^2} k_F^3 \tag{5.29}$$

$$= \frac{V}{3\pi^2} \left(\frac{2m_e E_F}{\hbar^2} \right)^{3/2} \tag{5.30}$$

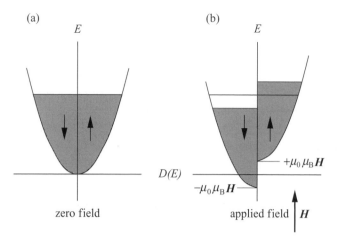

Figure 5.7 (a) Density of states in a free electron gas with no applied field. The up- and down-spin densities of states are equal and proportional to the square root of the energy. (b) Density of states in a free electron gas when a magnetic field is applied in the 'up' direction (i.e. parallel to the *down*-spin magnetic moments). The down-spin states (which have up magnetic moment) are lowered in energy, and the up-spin states are raised in energy, each by an amount $\mu_B H$.

where V is the volume of the crystal. Similarly, the number of electrons required to fill up the states to a general energy level E (below E_F) is $(V/3\pi^2)(2m_e E/\hbar^2)^{3/2}$. The density of states, $D(E)$, is defined as the derivative of the number of electron states with respect to energy. So differentiating the above expression gives us the density of states at the Fermi level,

$$D(E_F) = \frac{V}{2\pi^2}\left(\frac{2m_e}{\hbar^2}\right)^{3/2} E_F^{1/2}. \tag{5.31}$$

The number of electronic states per unit energy range is proportional to the square root of the energy, as shown in Fig. 5.7(a). We can simplify the expression by recognizing that $(V/3\pi^2)(2m_e/\hbar^2)^{3/2} = N/E_F^{3/2}$. Substituting gives

$$D(E_F) = \frac{3}{2}\frac{N}{E_F}. \tag{5.32}$$

Next let's use this expression for the density of states of a free electron gas to derive the susceptibility of our Pauli paramagnet.

5.4.3 Susceptibility of Pauli paramagnets

We saw in Chapter 3 that a single free electron, with spin angular momentum only, has a magnetic moment along the field direction of one Bohr magneton. (Remember, the moment along the field direction, $m = -g_e\mu_B m_s = \pm\mu_B$ for a free electron.)

Also, the application of a magnetic field will change the energy of the electron by an amount $\mu_0\mu_B \cos\theta$ where θ is the angle between the axis of the magnetic moment and the applied field. So an electron with its magnetic moment in the direction of the field will be lowered in energy by an amount $\mu_0\mu_B H$, and one antiparallel to the field will be increased in energy by $\mu_0\mu_B H$. Thus a magnetic field changes the density of states in a free electron gas as shown in Fig. 5.7(b).

If the field is applied in the up direction (so that it is parallel to the down-spin magnetic moment), there is a spill-over of electrons from up-spin to down-spin until the new Fermi levels for up- and down-spin are equal (and in fact very close to the original Fermi level, E_F.) The zero of energy for the down-spin density of states is at $-\mu_0\mu_B H$ and for the up-spin density of states at $+\mu_0\mu_B H$. Therefore the total number of down-spin electrons is now given by

$$\frac{1}{2}\int_{-\mu_0\mu_B H}^{E_F} D(E + \mu_0\mu_B H)\,dE \tag{5.33}$$

and of up-spin electrons

$$\frac{1}{2}\int_{+\mu_0\mu_B H}^{E_F} D(E - \mu_0\mu_B H)\,dE. \tag{5.34}$$

(The factor of $\frac{1}{2}$ occurs because only *one* electron occupies each up- or down-spin state, and the density of states was defined for two electrons per orbital).

The net magnetic moment, M, is the number of down-spin moments minus the number of up-spin moments, multiplied by the moment per spin, μ_B:

$$M = \frac{\mu_B}{2}\left[\int_{-\mu_0\mu_B H}^{E_F} D(E + \mu_0\mu_B H)\,dE - \int_{+\mu_0\mu_B H}^{E_F} D(E - \mu_0\mu_B H)\,dE\right]. \tag{5.35}$$

Changing variables gives

$$M = \frac{\mu_B}{2}\int_{E_F-\mu_0\mu_B H}^{E_F+\mu_0\mu_B H} D(E)\,dE. \tag{5.36}$$

The value of the integral is equal to the area of a strip of width $2\mu_0\mu_B H$ centered around E_F. This area is $2\mu_0\mu_B H D(E_F)$, so the net magnetic moment in the direction of the field is given by

$$M = \mu_0\mu_B^2 H D(E_F), \tag{5.37}$$

where $D(E_F)$ is the density of states at the Fermi level, which we derived earlier:

$$D(E_F) = \frac{3}{2}\frac{N}{E_F}. \tag{5.38}$$

So the susceptibility,

$$\chi = \frac{M}{H} = \frac{3N\mu_0\mu_B^2}{2E_F} \qquad (5.39)$$

which is independent of temperature! Remember that there is also a diamagnetic contribution to the susceptibility, which it turns out is one-third of the Pauli paramagnetism and of course in the opposite direction. Thus the expression for the total susceptibility of a metal which fits the free electron model is

$$\chi = \frac{\mu_0\mu_B^2 N}{E_F} \qquad (5.40)$$

(in SI units). The values of susceptibility calculated using this formula are in good agreement with measured values for metals such as Na or Al which are well described by the free electron model.

5.5 Paramagnetic oxygen

When two oxygen atoms (each with electronic configuration $1s^2$, $2s^2$, $2p^4$) join together to form an O_2 molecule, their atomic orbitals combine to form molecular orbitals, as shown in Fig. 5.8. (For an explanation of why the orbitals are ordered as shown, see Ref. 6.) The 16 electrons fill up the molecular orbitals from the lowest in energy up, and they occupy orbitals of equal energy individually before pairing up, just as they did in the atom. The consequence of this occupation scheme is that there are unpaired electrons in an O_2 molecule, and therefore gaseous oxygen has a paramagnetic response to an applied magnetic field.

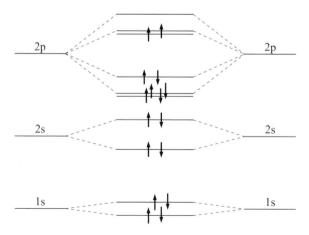

Figure 5.8 Molecular orbitals in oxygen.

5.6 Uses of paramagnets

Like the diamagnets, paramagnets do not find wide application because they have no permanent net magnetic moment. They are used, however, in the production of very low temperatures, by a process called adiabatic demagnetization. At a 'conventional' low temperature, such as that of liquid helium (a few degrees above absolute zero), the term α in the Langevin function is actually quite large – certainly greater than unity. Therefore if a paramagnet is cooled to liquid helium temperature in the presence of a strong magnetic field, the magnetization is nearly saturated, so most of the spins are lined up parallel to the field. If the paramagnet is then thermally isolated (for example by removing the liquid helium and leaving a good vacuum) and the field is turned off slowly, then the temperature of the paramagnet drops even further. The reason for the temperature drop is that, when the spins randomize as a result of the field's being removed, they must do work against whatever field remains. The only energy which is available to them is their thermal energy, and when they use this to demagnetize they lower their temperature. It is possible to reach temperatures as low as a few thousandths of a degree using this technique.

In addition, paramagnets allow us to study the electronic properties of materials which have atomic magnetic moments, without the interference of strong cooperative effects.

In the next chapter we will extend the Langevin theory of paramagnetism to help us start to understand the properties of the most important class of magnetic materials – the ferromagnets – in which the cooperative effects between magnetic moments are indeed strong.

Homework

Exercises

5.1   Show that the Brillouin function approaches the Langevin function as $J \to \infty$. What are the limits of the Brillouin function as $J \to \frac{1}{2}$ and $\alpha \to 0$?

5.2   Calculate the room temperature paramagnetic susceptibility of an ideal gas, in which each atom has $J = 1$ and $g = 2$. (Remember the ideal gas law: $PV = nRT$). These are in fact the values of J and g for molecular oxygen. Note that your answer is small and positive.

5.3   In this problem we will explore the properties of a model three-dimensional lattice of spins, each with spin, $S = \frac{1}{2}$.

  (a) What is the magnetic moment of each spin? What are the allowed values of the projection of the magnetic moment, m_i, onto some chosen axis, z say?
  (b) What are the possible values of the magnetic energy of each spin?
  (c) Assuming that the spins are non-interacting, calculate the magnetization of the lattice of spins when a magnetic field, H is applied along the z axis. (HINT: Use the

result from statistical thermodynamics that the average magnetization of a spin is given by $\langle M \rangle = (1/Z)\Sigma_i m_i e^{-E_i/k_B T}$ where m_i is the magnetization of a spin along the field direction when it has energy E_i, and $Z = \Sigma_i e^{-E_i/k_B T}$ is called the partition function.)

(d) For a given value of field, H, how does the magnetization, M, depend on temperature? Explain the behavior of M, for $T \to 0$. Taking the number of spins per unit volume to be 3.7×10^{28} m^{-3}, calculate the numerical value of the saturation magnetization, M_s, at $T = 0$. Explain the behavior of M for $T \to \infty$.

(e) What does the relationship between M and H reduce to for weak fields ($H \to 0$)? What is the expression for the susceptibility, χ, in this case, and how does it depend on temperature? Calculate the numerical value of χ at room temperature.

(f) Comment on the results which you have obtained for this spin system. What kind of magnetic behavior (antiferromagnetic, paramagnetic, diamagnetic, etc.) is displayed by this model system? Justify your conclusion. How would we need to modify the model in order to describe ferromagnetic behavior?

To think about

What mechanism might we use to lower the temperature below that obtained by the procedure described in Section 5.6?

Further reading

B.D. Cullity, *Introduction to magnetic materials.* Addison-Wesley, 1972, Chapter 3.

6

Interactions in ferromagnetic materials

"Anyone who is not shocked by quantum theory has not understood it."

Niels Bohr, 1885–1962

In Chapter 2 we introduced the concept of ferromagnetism, and looked at the hysteresis loop which characterizes the response of a ferromagnetic material to an applied magnetic field. This response is really quite remarkable! Look at Figs. 2.3 and 2.4 again – we see that it is possible to change the magnetization of a ferromagnetic material from an initial value of *zero*, to a saturation value of around 1000 emu/cm^3 by the application of a rather small magnetic field – around tens of oersteds.

The fact that the *initial* magnetization of a ferromagnet is zero is explained by the domain theory of ferromagnetism. The domain theory was postulated in 1907 by Weiss[18] and has been very successful. We will discuss the details of the domain theory, and the experimental evidence for the existence of domains, in the next chapter.

The subject of *this* chapter is: how can such a small external field cause such a large magnetization? In Exercise 6.2(b), you'll see that a field of 50 Oe has almost no effect on a system of *weakly interacting* elementary magnetic moments. Thermal agitations act to oppose the ordering influence of the applied field, and, when the atomic magnetic moments are independent, the thermal agitation wins. In ferromagnetic materials there must be a strong interaction between the magnetic moments, and we'll see later that this interaction is quantum mechanical in nature. We'll need to learn some more quantum mechanics as we go along, but hopefully we can make this as painless as possible.

But first let's start with the phenomenological model of ferromagnetism, which was again proposed by Weiss in his classic 1907 paper.[18] We won't worry about the *origin* of the strong interactions until Section 6.2 – instead we'll look first at their effect on observables such as susceptibility.

6.1 Weiss molecular field theory

In the previous chapter we showed that the Weiss molecular field explained the experimentally observed Curie–Weiss law for the behavior of many paramagnetic materials:

$$\chi = \frac{C}{T - \theta}. \tag{6.1}$$

Above their Curie temperatures, T_C, ferromagnetic materials become paramagnetic, and their susceptibilities follow the Curie–Weiss law, with a value of θ approximately equal to T_C. This experimental observation led Weiss to further assume that a molecular field acts in a ferromagnet *below* its Curie temperature as well as in the paramagnetic phase above T_C, and that this molecular field is strong enough to magnetize the substance even in the absence of an external applied field.

So we can regard a ferromagnetic material as being a paramagnet with a very large internal molecular field. This is a big help to us, because it means that we can use the theories of paramagnetism which we developed in the previous chapter to explain the properties of ferromagnets.

6.1.1 Spontaneous magnetization

First let's try to understand the spontaneous magnetization of ferromagnets using the Weiss theory. Remember that the classical Langevin theory of paramagnetism tells us that the magnetization is given by

$$\boldsymbol{M} = Nm L(\alpha), \tag{6.2}$$

where $\alpha = mH/k_B T$ and $L(\alpha)$ is the Langevin function. The solid line of Fig. 6.1 is a plot of $\boldsymbol{M} = Nm L(\alpha)$ as a function of α. But the Weiss theory gives us an additional expression for $\boldsymbol{M}$ – that $\boldsymbol{M} = \boldsymbol{H}_W/\gamma$, where γ is the molecular field constant. If we assume that the field $\boldsymbol{H}$ is provided entirely by the molecular field, then, since $\alpha = mH/k_B T$, the magnetization $\boldsymbol{M} = \boldsymbol{H}_W/\gamma$ is a linear function of α, plotted as the dashed line of Fig. 6.1. Then the only physical solutions are those where the two curves intersect. This occurs at the origin (which is unstable to any small fluctuation in the magnetization) and at the point $\boldsymbol{M}_{spont}$, where the material is spontaneously magnetized!

6.1.2 Effect of temperature on magnetization

We can also investigate the temperature dependence of the spontaneous magnetization using this graphical solution. Again if we assume that $\boldsymbol{H} = \boldsymbol{H}_W$, then

$$\alpha = \frac{mH_W}{k_B T} = \frac{m\gamma M}{k_B T}, \tag{6.3}$$

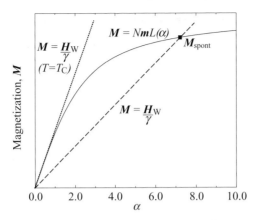

Figure 6.1 Explanation of spontaneous magnetization in ferromagnetic materials.

so

$$M = \left(\frac{k_B T}{m\gamma}\right)\alpha. \tag{6.4}$$

The magnetization is a linear function of α, with slope proportional to the temperature. So as the temperature increases, the slope of the dashed line increases, and it intersects the Langevin function at a point corresponding to a smaller spontaneous magnetization, M_{spont}.

Eventually, when the gradient of the dashed line approaches the initial tangent to the Langevin function, the spontaneous magnetization is zero. This is shown by the straight dotted line in Fig. 6.1. The temperature at this point is the Curie temperature, and at any higher temperature the *only* solution is at the origin, meaning that the spontaneous magnetization vanishes. The magnetization decreases smoothly to become zero at $T = T_C$, indicating (see Fig. 6.2) that the ferromagnetic to paramagnetic transition is a second-order phase transition.

The Curie temperature can be determined by equating, at the origin, the slope of the magnetization described by the Langevin function (which is $\frac{1}{3} \times Nm$), with the slope of the straight line representing magnetization by the molecular field:

$$\frac{k_B T_C}{m\gamma} = \frac{1}{3} \times Nm, \tag{6.5}$$

so

$$T_C = \frac{\gamma Nm^2}{3k_B}. \tag{6.6}$$

A large molecular field constant leads to a high Curie temperature. This is what we would expect intuitively – magnetic moments which interact strongly with each

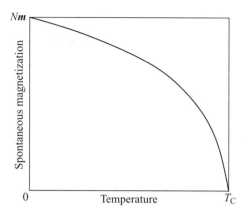

Figure 6.2 Variation of spontaneous magnetization with temperature in ferromagnetic materials, calculated using classical Langevin theory.

other should require a larger thermal energy to disrupt their magnetic ordering and induce a phase transition to a paramagnetic phase.

Conversely, if the Curie temperature is known, then the Weiss molecular field can be extracted:

$$\gamma = \frac{3k_B T_C}{Nm^2} \tag{6.7}$$

and

$$H_W = \gamma M = \gamma Nm = \frac{3k_B T_C}{m}. \tag{6.8}$$

(Note that this is similar to the approximate expression we obtained earlier simply by equating the magnetic dipole energy with the thermal energy).

A schematic plot of spontaneous magnetization versus temperature obtained using this graphical technique is given in Fig. 6.2. Such plots reproduce experimental results reasonably well. Greater accuracy can be obtained by substituting the Langevin function by the quantum mechanical Brillouin function, with an appropriate choice of J. Using the quantum mechanical expressions, the molecular field constant is given by

$$\gamma = \frac{3k_B T_C}{Nm_{eff}^2} \tag{6.9}$$

and the Curie temperature by

$$T_C = \frac{\gamma Nm_{eff}^2}{3k_B}. \tag{6.10}$$

with $m_{eff} = g\sqrt{J(J+1)}\mu_B$. Multiplication by μ_0 produces the expression for T_C in SI units.

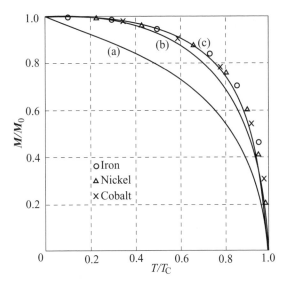

Figure 6.3 Relative spontaneous magnetization of Fe, Co and Ni as a function of relative temperature. The curves are calculated using either the classical Langevin function (a) or the Brillouin function with $J = 1$ (b) and $J = \frac{1}{2}$ (c). From Ref. 19, F. Tyler, *Phil. Mag.* **11** 596. Copyright (1931) Taylor & Francis Ltd, (website:- http://www.tandf.co.uk/journals).

Figure 6.3 compares the measured relative spontaneous magnetizations of Fe, Co and Ni as a function of temperature, with curves predicted using classical Langevin theory, and using the Brillouin function with $J = 1$ and $J = \frac{1}{2}$. It is clear that the Weiss theory, with $J = 1$, gives good agreement with experiment.

6.2 Origin of the Weiss molecular field

In 1928, Heisenberg showed that the existence of a Weiss "molecular field" could be explained using a quantum mechanical treatment of the many-body problem.[20] In the next section we will work through the quantum mechanical calculation for the energy of the helium atom, which has two electrons and therefore provides a simple example of a many-body Hamiltonian. The relevant result which emerges from the quantum mechanics is that there is a term of electrostatic origin in the energy of interaction between neighboring atoms which tends to orient the electron spins parallel to each other. This term is called the *exchange* integral, and it does not have a classical analog.

The exchange interaction is in fact a consequence of the Pauli exclusion principle. If two electrons in an atom have antiparallel spins, then they are allowed to share the same atomic or molecular orbital. As a result they will overlap spatially, thus increasing the electrostatic Coulomb repulsion. In contrast, if they have parallel spins, then they must occupy different orbitals and so will have *less* unfavorable

Coulomb repulsion. (This is the same argument that we used to explain Hund's first rule in Chapter 3.) So the orientation of the spins affects the spatial part of the wavefunction, and this in turn determines the electrostatic Coulomb interaction between the electrons.

Let's make a rough classical estimate of the order of magnitude of the Coulomb repulsion between two electrons. If we assume that the average distance between electrons is around 1 Å then the Coulomb energy,

$$U = \frac{e^2}{4\pi\epsilon_0 r} \approx \frac{(1.6 \times 10^{-19})^2}{(1.1 \times 10^{-10})(1 \times 10^{-10})} J \approx 2.1 \times 10^{-18} \, J = 1.4 \times 10^5 \, K.$$

(6.11)

This is about 10^5 times larger than the magnetic dipolar interaction which we calculated in Exercise 1.3c. So if the electron distribution is changed even by a small amount, the effect on the total energy of an atom can be significant. This explains why the effective molecular field is so large!

6.2.1 Quantum mechanics of the He atom

Now let's calculate the energies for the excited state of helium where one electron is in the 1s atomic orbital and one is in the 2s, for both parallel and antiparallel spin arrangements, as shown in Fig. 6.4. (We can't use the ground state $1s^2$ configuration as an illustration because here the two electrons can *only* exist in the antiparallel configuration). We can write the electronic Hamiltonian, H, as the sum of three terms – one term for each of the electrons interacting with the nucleus, plus one for the interaction between the electrons:

$$H = H_1 + H_2 + H_{12}.$$

(6.12)

Here H_1 and H_2 consist of a kinetic energy part plus the Coulomb energy between each electron and the nucleus, and H_{12} is the Coulomb interaction between the two

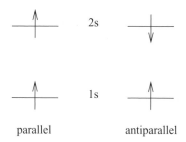

Figure 6.4 Parallel and antiparallel arrangements of spins in the He first excited state.

electrons.

$$H_1 = -\frac{\hbar^2}{2m_e}\nabla_1^2 - \frac{Ze^2}{4\pi\epsilon_0 r_1},$$ (6.13)

$$H_2 = -\frac{\hbar^2}{2m_e}\nabla_2^2 - \frac{Ze^2}{4\pi\epsilon_0 r_2},$$ (6.14)

$$H_{12} = \frac{e^2}{4\pi\epsilon_0 r_{12}}.$$ (6.15)

Here r_{12} is the distance between the electrons, and Z is the atomic number.

We could solve the Schrödinger equation for this Hamiltonian using quantum mechanical perturbation theory. (If you'd like to see the full derivation, there is a nice discussion in the book by Atkins.[6]) However, to avoid being overwhelmed with quantum mechanics and losing the thread of the argument, instead we'll just write down the form of the wavefunctions for the cases with spins aligned both parallel and antiparallel, using the Pauli exclusion principle to guide us. We need to use the full statement of the Pauli principle, that the total electronic wavefunction of a system must be antisymmetric with respect to the interchange of two electrons. We can make a simplistic argument for why this is the case by thinking about our two-electron atom example – if two electrons occupy the same molecular orbital, then interchanging them will have no effect on the spatial part of the wavefunction. However in order to occupy the same molecular orbital they must have opposite spins. So switching the electrons will change the sign of the spin part. The total wavefunction is the product of spin and spatial parts, and that product always ends up with the opposite sign.

Any state which is antisymmetric for the interchange of spin coordinates (i.e. in which the spins are aligned antiparallel) is always symmetric for the interchange of spatial coordinates. A molecular wavefunction for He which satisfies the spatial symmetry criterion has the form

$$\Psi(\mathbf{r}_1, \mathbf{r}_2) = \frac{1}{\sqrt{2}}\left[\phi_{1s}(\mathbf{r}_1)\phi_{2s}(\mathbf{r}_2) + \phi_{2s}(\mathbf{r}_1)\phi_{1s}(\mathbf{r}_2)\right],$$ (6.16)

where ϕ_{1s} and ϕ_{2s} are the 1s and 2s atomic orbitals and $\mathbf{r}_1$ and $\mathbf{r}_2$ are the positions of electrons 1 and 2. (The $1/\sqrt{2}$ is for normalization.) Similarly, a state which is symmetric for the interchange of spin coordinates (i.e. in which the spins are aligned parallel) must be antisymmetric for the interchange of spatial coordinates. A molecular wavefunction which satisfies this criterion has the form

$$\Psi(\mathbf{r}_1, \mathbf{r}_2) = \frac{1}{\sqrt{2}}\left[\phi_{1s}(\mathbf{r}_1)\phi_{2s}(\mathbf{r}_2) - \phi_{2s}(\mathbf{r}_1)\phi_{1s}(\mathbf{r}_2)\right].$$ (6.17)

(If we had worked through all the quantum mechanics we would in fact have found three degenerate solutions with spatially symmetric wavefunctions, and one with the spatially antisymmetric wavefunction.)

Now let's calculate the energy of each of these states using the Hamiltonian of Eqn 6.12. Using Dirac bra-ket notation, the total energy, E, is

$$
\begin{aligned}
E &= \langle \Psi(\mathbf{r}_1, \mathbf{r}_2)|H|\Psi(\mathbf{r}_1, \mathbf{r}_2)\rangle \\
&= \tfrac{1}{2}\langle[\phi_{1s}(\mathbf{r}_1)\phi_{2s}(\mathbf{r}_2) \pm \phi_{2s}(\mathbf{r}_1)\phi_{1s}(\mathbf{r}_2)] \\
&\quad |(H_1 + H_2 + H_{12})| [\phi_{1s}(\mathbf{r}_1)\phi_{2s}(\mathbf{r}_2) \pm \phi_{2s}(\mathbf{r}_1)\phi_{1s}(\mathbf{r}_2)]\rangle \\
&= \tfrac{1}{2}[\langle\phi_{1s}(\mathbf{r}_1)|H_1|\phi_{1s}(\mathbf{r}_1)\rangle + \langle\phi_{2s}(\mathbf{r}_1)|H_1|\phi_{2s}(\mathbf{r}_1)\rangle \\
&\quad + \langle\phi_{1s}(\mathbf{r}_2)|H_2|\phi_{1s}(\mathbf{r}_2)\rangle + \langle\phi_{2s}(\mathbf{r}_2)|H_2|\phi_{2s}(\mathbf{r}_2)\rangle \\
&\quad + \langle\phi_{1s}(\mathbf{r}_1)\phi_{2s}(\mathbf{r}_2)|H_{12}|\phi_{1s}(\mathbf{r}_1)\phi_{2s}(\mathbf{r}_2)\rangle \\
&\quad + \langle\phi_{2s}(\mathbf{r}_1)\phi_{1s}(\mathbf{r}_2)|H_{12}|\phi_{2s}(\mathbf{r}_1)\phi_{1s}(\mathbf{r}_2)\rangle \\
&\quad \pm \langle\phi_{1s}(\mathbf{r}_1)\phi_{2s}(\mathbf{r}_2)|H_{12}|\phi_{2s}(\mathbf{r}_1)\phi_{1s}(\mathbf{r}_2)\rangle \\
&\quad \pm \langle\phi_{2s}(\mathbf{r}_1)\phi_{1s}(\mathbf{r}_2)|H_{12}|\phi_{1s}(\mathbf{r}_1)\phi_{2s}(\mathbf{r}_2)\rangle] \\
&= E_1 + E_2 + K \pm J, \qquad \text{say.}
\end{aligned}
\tag{6.18}
$$

Remember that the $+$ sign corresponds to antiparallel spins, and the $-$ sign to parallel spins. We see that the energy for parallel orientation of the spins is less than the energy for antiparallel orientation by an amount $2J$ when J is positive. So a positive J favors parallel spins, which corresponds to ferromagnetic ordering! Here

$$
E_1 = \langle\phi_{1s}(\mathbf{r}_1)|H_1|\phi_{1s}(\mathbf{r}_1)\rangle = \langle\phi_{1s}(\mathbf{r}_2)|H_2|\phi_{1s}(\mathbf{r}_2)\rangle \tag{6.19}
$$

$$
E_2 = \langle\phi_{2s}(\mathbf{r}_1)|H_1|\phi_{2s}(\mathbf{r}_1)\rangle = \langle\phi_{2s}(\mathbf{r}_2)|H_2|\phi_{2s}(\mathbf{r}_2)\rangle \tag{6.20}
$$

$$
\begin{aligned}
K &= \langle\phi_{1s}(\mathbf{r}_1)\phi_{2s}(\mathbf{r}_2)|H_{12}|\phi_{1s}(\mathbf{r}_1)\phi_{2s}(\mathbf{r}_2)\rangle \\
&= \langle\phi_{2s}(\mathbf{r}_1)\phi_{1s}(\mathbf{r}_2)|H_{12}|\phi_{2s}(\mathbf{r}_1)\phi_{1s}(\mathbf{r}_2)\rangle
\end{aligned}
\tag{6.21}
$$

$$
\begin{aligned}
J &= \langle\phi_{1s}(\mathbf{r}_1)\phi_{2s}(\mathbf{r}_2)|H_{12}|\phi_{2s}(\mathbf{r}_1)\phi_{1s}(\mathbf{r}_2)\rangle \\
&= \langle\phi_{2s}(\mathbf{r}_1)\phi_{1s}(\mathbf{r}_2)|H_{12}|\phi_{1s}(\mathbf{r}_1)\phi_{2s}(\mathbf{r}_2)\rangle.
\end{aligned}
\tag{6.22}
$$

Here E_1 and E_2 represent the energies of the 1s and 2s orbitals respectively in the field of the helium nucleus, K is the Coulomb interaction between the electron densities ϕ_{1s}^2 and ϕ_{2s}^2, and J is the exchange interaction, which clearly has no classical analog.

6.3 Collective-electron theory of ferromagnetism

We have seen that Weiss's idea of the molecular field, combined with the Langevin theory of localized moments, gives a rather good description of many properties of ferromagnetic materials. The temperature dependence of the spontaneous magnetization compares favorably with the observed values, and the existence of a phase transition to a paramagnetic state is explained. However the localized-moment theory breaks down in one important aspect – it is unable to account for the measured values of the magnetic moment per atom in some ferromagnetic materials, particularly in ferromagnetic metals. There are two significant discrepancies. First, according to the Weiss theory, the magnetic dipole moment on each atom or ion should be the same in both the ferromagnetic and paramagnetic phases. Experimentally this is not the case. Second, in the localized-moment theory, the magnetic dipole moment on each atom or ion should correspond to an integer number of electrons. Again this is not observed experimentally. To explain the data we need to use the band theory, or collective-electron theory, which we introduced earlier in our discussion of Pauli paramagnetism.

The mechanism producing magnetism in ferromagnetic metals is ultimately the same exchange energy that gives rise to Hund's rules in atoms. This exchange energy is minimized if all the electrons have the same spin. Opposing the alignment of spins is the increased band energy involved in transferring electrons from the lowest band states (occupied with up- and down-spin electrons) to band states of higher energy. This band energy prevents simple metals from being ferromagnetic.

In the elemental ferromagnetic transition metals, Fe, Ni and Co, the Fermi energy lies in a region of overlapping 3d and 4s bands, as shown schematically in Fig. 6.5. We will assume that the structures of the 3d and 4s bands do not change markedly across the first transition series, and so any differences in electronic structure are

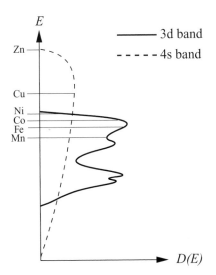

Figure 6.5 Schematic 3d and 4s densities of states in transition metals. The positions of the Fermi levels in Zn, Cu, Ni, Co, Fe and Mn are shown.

caused entirely by changes in the Fermi energy. This approximation is known as the *rigid-band model*, and detailed band structure calculations have shown that it is a reasonable assumption.

As a result of the overlap between the 4s and 3d bands, the valence electrons partially occupy the 3d and 4s band. For example, Ni, with 10 valence electrons per atom, has 9.46 electrons in the 3d band and 0.54 electrons in the 4s band. The 4s band is broad, with a low density of states at the Fermi level. Consequently, the energy which would be required to promote a 4s electron into a vacant state so that it could reverse its spin is more than that which would be gained by the resulting decrease in exchange energy. By contrast, the 3d band is narrow and has a much higher density of states at the Fermi level. The large number of electrons near the Fermi level reduces the band energy required to reverse a spin, and the exchange effect dominates.

It is useful to picture the exchange interaction as shifting the energy of the 3d band for electrons with one spin direction relative to the band for electrons with the opposite spin direction. The magnitude of the shift is independent of the wavevector, giving a rigid displacement of the states in a band with one spin direction relative to the states with the opposite spin direction. If the Fermi energy lies within the 3d band, then the displacement will lead to more electrons of the lower-energy spin direction and hence a spontaneous magnetic moment in the ground state. The resulting band structure looks similar to that of a Pauli paramagnet in an external magnetic field. The difference is that in this case the *exchange interaction* causes the change in energy, and an external field is not required to induce the magnetization.

Figure 6.6 shows the 4s and 3d densities of states within this picture. The exchange splitting is negligible for the 4s electrons, but significant for 3d electrons. In Ni, for example, the exchange interaction displacement is so strong that one 3d sub-band is filled with 5 electrons, and the other contains all 0.54 holes. So the saturation magnetization of Ni is $M_s = 0.54N\mu_B$, where N is the total number of Ni atoms in the sample. We now see why the magnetic moments of the transition metals do not correspond to integer numbers of electrons! This model also explains why the later transition metals, Cu and Zn, are not ferromagnetic. In Cu, the Fermi level lies *above* the 3d band. Since both the 3d sub-bands are filled, and the 4s band has no exchange-splitting, then the numbers of up- and down-spin electrons are equal. In Zn, both the 3d and 4s bands are filled and so do not contribute a magnetic moment.

For the lighter transition metals, Mn, Cr, etc., the exchange interaction is less strong, and the band energy is larger. So the energy balance is such that ferromagnetism is not observed. In fact both Mn and Cr actually have rather complicated spin arrangements which are antiferromagnetic in nature. More about that later!

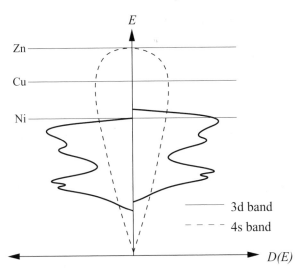

Figure 6.6 Schematic 3d and 4s up- and down-spin densities of states in a transition metal with exchange interaction included.

6.3.1 The Slater–Pauling curve

The collective-electron and rigid-band models are further supported by the rather well-known plot known as the Slater–Pauling curve.[21, 22] In the late 1930s, both Slater and Pauling independently calculated the saturation magnetization as a continuous function of the number of 3d and 4s valence electrons per atom across the first transition series. They used a rigid-band model, and obtained a linear increase in saturation magnetization from Cr to Fe, then a linear decrease, reaching zero magnetization at an electron density between Ni and Cu. They compared their calculated values with measured magnetizations of the pure ferromagnets Fe, Co and Ni, as well as Fe–Co, Co–Ni and Ni–Cu alloys. The results from Pauling's paper are shown in Fig. 6.7. The measured values agree well with the theoretical values. Although there are only three *pure* ferromagnetic metals, many transition metal alloys are ferromagnetic, and the saturation magnetic moment is more or less linearly dependent on the number of valence electrons.

6.4 Summary

In this chapter (and in the previous chapter on paramagnetism) we have introduced and applied two complementary theories of magnetism. In the localized moment theory, the valence electrons are attached to the atoms and cannot move about the

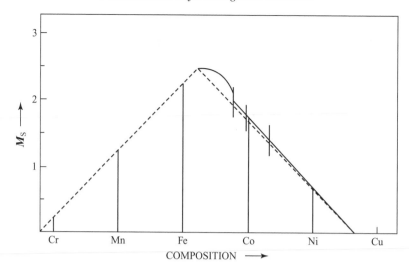

Figure 6.7 Comparison of experimental values (solid curves) and predicted values (dashed lines) of saturation magnetic moment per atom (M_s) for Fe–Co, Co–Ni and Ni–Cu alloys. The short vertical lines indicate change in crystal structure. Cr, Mn and their alloys do not exhibit ferromagnetism (see text). From Ref. 22. Copyright (1938) by the American Physical Society.

crystal. The valence electrons contribute a magnetic moment which is localized at the atom. The localized moment theory accounts for the variation of spontaneous magnetization with temperature in the ferromagnetic phase, and explains the Curie–Weiss behavior above the Curie temperature. In the collective-electron model, or band theory, the electrons responsible for magnetic effects are ionized from the atoms, and are able to move through the crystal. Band theory explains the non-integer values of the magnetic moment per atom that are observed in metallic ferromagnets.

Of course in 'real life' neither model is really correct, although there are some materials for which one or the other is a rather good approximation. In the rare earth elements and their alloys, for example, magnetism comes from the tightly bound core f electrons, and so the localized moment model works well. In materials such as Ni_3Al the electrons are highly itinerant, so the band theory gives accurate results. Transition metals show some features of both localization and of itinerant electrons. Permanent magnets, such as $NdFe_{14}B$, are particularly hard to describe since they combine the behavior of both transition metals and rare earths.

By far the most successful method currently available for calculating the magnetic properties of solids is density functional theory (DFT). DFT is an *ab initio* many-body theory which includes (in principle) *all* the interactions between *all* the electrons. No assumptions are made as to whether the electrons are localized or itinerant – rather the electrons choose the arrangement which will give them the lowest possible total energy. Unfortunately, DFT calculations are both computationally intensive and difficult, in particular because the exact form of the exchange and correlation part of the inter-electronic interaction energy is not known. As an example, it has only recently been possible to obtain the correct body-centered cubic, ferromagnetic ground state for iron.[23] (Earlier studies predicted that it should be non-magnetic and face-centered cubic!) An excellent review of the use of DFT to calculate the properties of magnetic materials is given by Jansen.[24]

Homework

Exercises

6.1   The Curie temperature of nickel, $T_C = 628.3$ K, and the saturation magnetization is $0.6\mu_B$ per atom. Calculate the molecular field according to the Weiss theory. Your answer should be very large!

6.2   In Exercise 1.3 you calculated the field generated by an electron circulating in a 1 Å radius orbit, at a distance of 3 Å from the center of the orbit. These values are typical for an atom in a transition metal crystal.

(a) To what Curie temperature would this field correspond?

(b) You also calculated the dipole moment of the electron. To what extent would your calculated magnetic moment be affected by an external field of say 50 Oe? (Use $E = -\boldsymbol{m} \cdot \boldsymbol{H}$ and compare the result with the thermal energy, $k_B T$, at room temperature.)

6.3   Review question

(a) Using Ampère's circuital law or the Biot–Savart law, make an order-of-magnitude estimate of the size of the magnetic field generated by the valence electrons in a Ni atom at a distance corresponding to the Ni–Ni spacing in a solid sample of Ni. (Assume that the field arises from the current generated by the circulation of *unpaired* electrons around the nucleus.)

(b) Use Hund's rules to determine the values of S, L and J for an isolated Ni atom with electronic structure $(4s)^2(3d)^8$. What are the allowed values of the magnetic moment along the field axis for a Ni atom?

(c) Use your answers to parts (a) and (b) to estimate the difference in magnetic dipole energy between Ni atoms aligned parallel and antiparallel to each other.

(d) Given that the Curie temperature of Ni is $358\,°$C, how does the magnetic dipole energy which you calculated in (c) compare with the actual strength of the ferromagnetic coupling between Ni atoms?

(e) Explain briefly (a few sentences) what the true origin of the ferromagnetic coupling is in Ni.

(f) The actual value of the magnetic dipole moment in metallic Ni is $0.54\mu_B$. Why does this number *not* correspond to an integer number of electrons? (You'll probably need a diagram to explain this one!)

Further reading

B.D. Cullity, *Introduction to magnetic materials*. Addison-Wesley, 1972, Chapter 4.
H.J.F. Jansen, *Physics Today* (April) 50, 1995.

7

Ferromagnetic domains

"O care! O guilt! – O vales and plains,
Here, 'mid his own unvexed domains,
A Genius dwells,"
William Wordsworth, from "The pass of Kirkstone", *The complete
poetical works.* Macmillan and Co., London, 1888.

Ferromagnetic domains are small regions in ferromagnetic materials within which all the magnetic dipoles are aligned parallel to each other. When a ferromagnetic material is in its demagnetized state, the magnetization vectors in different domains have different orientations, and the total magnetization averages to zero. The process of magnetization causes all the domains to orient in the same direction. The purpose of this chapter is to explain why domains occur, to describe their structure, and the structure of their boundaries, and to discuss how they affect the properties of materials. As a preliminary, we will describe some experiments which allow us to observe domains directly with rather simple equipment.

7.1 Observing domains

Domains are usually too small to be seen using the naked eye. Fortunately there are a number of rather straightforward methods for observing them. The first method was developed by Francis Bitter in 1931.[25] In the Bitter method, the surface of the sample is covered with an aqueous solution of very small colloidal particles of magnetite, Fe_3O_4. The magnetite deposits as a band along the domain boundaries at their intersection with the sample surface. The outlines of the domains can then be seen using a microscope. Figure 7.1 is taken from Bitter's original 1931 publication; the light-colored lines are magnetite deposits on a crystal of nickel at 16 times magnification.

As we will discuss later in the chapter, at the domain boundaries the directions of the magnetic dipole moments change, and poles are formed at the surface of the

Figure 7.1 Magnetite deposits (light-colored lines) on a crystal of nickel. Width of field 3.125 mm. From Ref. 25. Copyright (1931) by the American Physical Society.

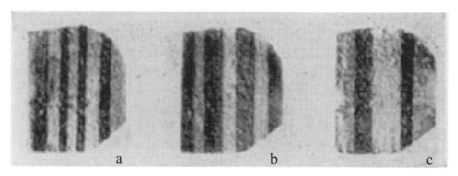

Figure 7.2 Three different domain configurations in a demagnetized sample of silicon iron. Width of each sample ~10 mm. The crystal was demagnetized in each case by an alternating field of decreasing amplitude, and the duration of the demagnetizing process decreased from (a) to (b) to (c). From Ref. 27. Copyright (1954) by the American Physical Society.

sample. A magnetic field originates at the pole, and this attracts the fine magnetic particles to it. Note therefore that the Bitter method actually observes the domain boundaries, rather than the domains themselves. The technique can also be used to observe domain wall motion, because the magnetite particles follow the intersection of the wall with the surface. However the sample must first be carefully cleaned and polished so that the magnetite particles don't get stuck in cracks or around impurities.

It is also possible to observe domains using polarized light. As a result of the magneto-optic effect (which we will discuss in detail in Chapter 12), the plane of polarized light is rotated when it either passes through, or is reflected from, magnetic material. The direction of rotation depends on the orientation of magnetization. Therefore regions of the sample with opposite orientations of the magnetization will rotate the polarized light in opposite directions. This method was first used in the early 1950s;[26] in Fig. 7.2 we show photographs of domains in demagnetized silicon iron from an early application of the technique.[27]

Note that both the Bitter and magneto-optic techniques are sensitive to the domain structure at the surface of the sample. The surface domain structure is sensitive to local details of flux closure on the surface, and can be more complicated than the basic domain structure running through the bulk of the sample.

7.2 Why domains occur

We saw in the previous chapter that quantum mechanics gives rise to an exchange energy which tends to align electron spins, and hence their magnetic dipole moments, parallel to each other. The exchange energy provides a strong driving force for parallel alignment, therefore we might expect that ferromagnetic materials should be composed of one single domain, with all dipoles aligned in the same direction.

Although a single domain would certainly minimize the *exchange* contribution to the total energy, there are a number of other contributions to the total magnetic energy of a ferromagnet. The formation of domains allows a ferromagnetic material to minimize its *total* magnetic energy, of which the exchange energy is just one component. The other main contributors to the magnetic energy are the magnetostatic energy, which is the principal driving force for domain formation, and the magnetocrystalline and magnetostrictive energies, which influence the shape and size of domains. Next we will discuss each of these energy contributions in turn, and show how they determine the formation and structure of domains in ferromagnetic materials.

7.2.1 Magnetostatic energy

A magnetized block of ferromagnetic material containing a single domain has a macroscopic magnetization. The magnetization causes the block to behave as a magnet, with a magnetic field around it. Figure 7.3(a) illustrates a magnetized block with its associated external field. It is apparent from the figure that the field acts to magnetize the block in the *opposite* direction from its own magnetization. For this reason it is called the *demagnetizing field*, H_d. We will encounter demagnetizing fields again in Chapter 10 when we discuss shape anisotropy.

The demagnetizing field causes a magnetostatic energy which depends on the shape of the sample. It is this magnetostatic energy which allows the block to

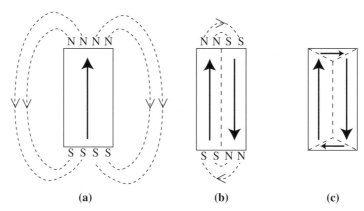

Figure 7.3 Reduction of the magnetostatic energy by domain formation in a ferromagnet.

do work such as lifting another ferromagnet against the force of gravity. The magnetostatic energy can be reduced by reducing the external demagnetizing field – one way to do this is to divide the block into domains, as shown in Fig. 7.3(b). Here the external field is lower, so that the block is capable of doing less work, and (conversely) is storing less magnetostatic energy. Of course the magnetic moments at the boundary between the two domains are not able to align parallel, so the formation of domains increases the exchange energy of the block.

To reduce the magnetostatic energy to zero, we need a domain pattern which leaves no magnetic poles at the surface of the block. One way to achieve this is shown in Fig. 7.3(c). Before we can decide whether this is a likely domain pattern, we need to understand a little about the magnetocrystalline and magnetostrictive energy contributions.

7.2.2 Magnetocrystalline energy

The magnetization in ferromagnetic crystals tends to align along certain preferred crystallographic directions. The preferred directions are called the 'easy' axes, since it is easiest to magnetize a demagnetized sample to saturation if the external field is applied along a preferred direction. Figure 7.4 shows schematic magnetization curves for a ferromagnetic single crystal, with the field applied along the easy and hard axes. In both cases the same saturation magnetization is achieved, but a much larger applied field is required to reach saturation along the hard axis than along the easy axis.

Different materials have different easy axes – in body-centered cubic (bcc) iron the easy axis is the $\langle 100 \rangle$ direction (the cube edge). Of course since bcc iron is a cubic crystal, all six cube edge orientations ($\langle 100 \rangle$, $\langle 010 \rangle$, $\langle 001 \rangle$, $\langle \bar{1}00 \rangle$, $\langle 0\bar{1}0 \rangle$ and $\langle 00\bar{1} \rangle$) are in fact equivalent easy axes. The body diagonal is the hard axis of

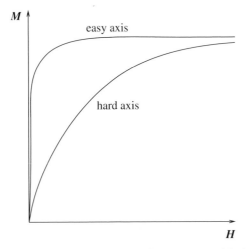

Figure 7.4 Schematic magnetization curves for a ferromagnet with the field oriented along the hard and easy directions.

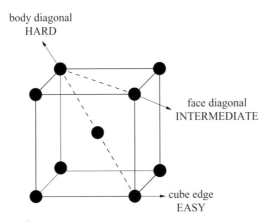

Figure 7.5 Easy, medium and hard directions of magnetization in a unit cell of bcc iron.

magnetization, and other orientations, such as the face diagonal are intermediate. A unit cell of iron, with the easy, medium and hard directions of magnetization labeled is shown in Fig. 7.5.

By contrast, the easy axis of face-centered cubic (fcc) Ni is the ⟨111⟩ body diagonal, and of hexagonal close-packed (hcp) Co it is the ⟨0001⟩ direction.

The phenomenon that causes the magnetization to align itself along a preferred crystallographic direction is the magnetocrystalline anisotropy. The crystal is higher in energy when the magnetization points along the hard direction than along the easy direction, and the energy difference between samples magnetized along easy and hard directions is called the magnetocrystalline anisotropy energy. In fact the area between the hard and easy magnetization curves in Fig. 7.4 is a measure of the magnetocrystalline energy for that material. We will discuss the details of magnetocrystalline anisotropy, including its physical origin, how it is measured and why it is useful, in Chapter 10. For now we are interested in how it affects the structure of magnetic domains.

To minimize the magnetocrystalline energy, domains will form so that their magnetizations point along easy crystallographic directions. So, for example, the 'vertical' axis in Fig. 7.3 should correspond to a cube edge in bcc iron. Because of the cubic symmetry, the horizontal direction is also an easy axis for bcc iron, therefore the domain arrangement shown in Fig. 7.3(c) has a low magnetocrystalline energy.

The horizontal domains at the top and bottom of the crystal in Fig. 7.3(c) are called 'domains of closure', and they form readily when a material has easy axes perpendicular to each other. In such materials, this configuration is particularly favorable because it eliminates the demagnetizing field, and hence the magnetostatic energy, without increasing the magnetocrystalline anisotropy energy. However an additional energy, called the magnetostrictive energy, is introduced; we discuss this next.

One more point to note is that the magnetocrystalline energy clearly affects the structure of the domain boundaries. Within the region between domains the direction of magnetization changes, and therefore cannot be aligned along an easy direction. So, like the exchange energy, the magnetocrystalline energy prefers large domains with few boundaries.

7.2.3 Magnetostrictive energy

When a ferromagnetic material is magnetized it undergoes a change in length known as its magnetostriction. Some materials, such as iron, elongate along the direction of magnetization and are said to have a positive magnetostriction. Others, such as nickel, contract and have negative magnetostriction. The length changes are very small – tens of parts per million – but do influence the domain structure.

In iron, magnetostriction causes the triangular domains of closure to try to elongate horizontally, whereas the long vertical domains try to elongate vertically, as shown in Fig. 7.6. Clearly the horizontal and vertical domains can't elongate at the same time, and instead an elastic strain energy term is added to the total energy. The elastic energy is proportional to the volume of the domains of closure, and can be lowered by reducing the size of the closure domains, which in turn requires smaller primary domains. Of course making smaller domains introduces additional domain walls, and the corresponding increase in exchange and magnetostatic energy. The total energy is reduced by a compromise domain arrangement such as that shown in Fig. 7.7.

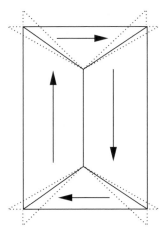

Figure 7.6 Magnetostriction in the triangular domains of closure in bcc iron.

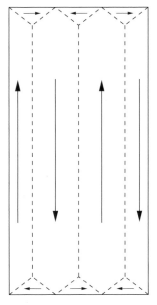

Figure 7.7 A domain arrangement that reduces the sum of the exchange, magnetostatic, magnetocrystalline, and domain wall energies to a minimum.

7.3 Domain walls

The boundaries between adjacent domains in bulk ferromagnetic materials are called domain walls, or Bloch walls. They are about four-millionths of an inch ($\sim$10 μm) in thickness, and across this distance the direction of magnetization changes usually by either 180 or 90 degrees.

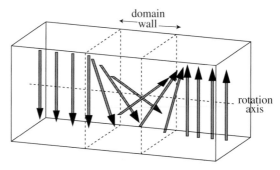

Figure 7.8 Change in orientation of the magnetic dipoles in a 180° twist boundary.

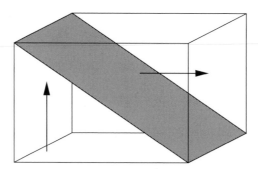

Figure 7.9 Change in orientation of the magnetic dipoles in a 90° tilt boundary.

The width of domain walls is again determined by a balance between competing energy contributions. The exchange energy is optimized if adjacent magnetic moments are parallel, or as close to parallel as possible, to each other. This favors wide walls, so that the change in angle of the moments between adjacent planes of atoms can be as small as possible. However the magnetocrystalline anisotropy is optimized if the moments are aligned as closely as possible to the easy axes. This favors narrow walls with a sharp transition between the domains, so that few moments have unfavorable crystalline alignment in the transition region. In practice a compromise is reached which minimizes the total energy across the boundary.

The most energetically favorable types of domain wall are those which do not produce magnetic poles within the material, and therefore don't introduce demagnetizing fields. One such wall is the twist boundary, illustrated for a 180° boundary in Fig. 7.8. Here the magnetization perpendicular to the boundary does not change across the wall, therefore no magnetic poles or demagnetizing fields arise. Also stable are 90° tilt boundaries, as shown in Fig. 7.9. The magnetic moments rotate through the wall in such a way that they make a constant angle of 45° with both the wall normal and the surface.

Figure 7.10 Rotation of the spins in a Néel wall.

Another kind of domain wall, called a Néel wall, occurs in thin films of magnetic materials. In Néel walls the spins rotate around an axis normal to the surface of the film, rather than around an axis normal to the domain wall. The spin rotation in a Néel wall is shown in plan view in Fig. 7.10. Néel walls are energetically favorable in thin films because free poles are formed on the wall surface, rather than the film surface, causing a reduction in magnetostatic energy.

7.4 Magnetization and hysteresis

Now that we understand a little about the structure and origin of domains, let's look at how they influence the magnetization and hysteresis curves of ferromagnetic materials. Figure 7.11 shows a schematic magnetization curve for a ferromagnetic material, with a sketch of the domain structure at each stage of the magnetization. The magnetic field is applied at an angle (horizontal in the picture) which is slightly off the easy axis of magnetization. In the initial demagnetized state, the domains are arranged such that the magnetization averages to zero. When the field is applied, the domain whose magnetization is closest to the field direction starts to grow at the expense of the other domains. The growth occurs by domain wall motion. At first the domain wall motion is reversible; if the field is removed during the reversible stage, the magnetization retraces its path and the demagnetized state is regained. In this region of the magnetization curve the sample does not show hysteresis.

After a while, the moving domain walls encounter imperfections such as defects or dislocations in the crystal. Crystal imperfections have an associated magnetostatic energy. However when a domain boundary intersects the imperfection, this magnetostatic energy can be eliminated, as shown in Fig. 7.12. The intersection of the domain boundary with the imperfection is a local energy minimum. As a result the domain boundary will tend to stay pinned at the imperfection, and energy is required to move it past the imperfection. This energy is provided by the external magnetic field. A typical variation of Bloch wall energy with position in an imperfect crystal is shown in Fig. 7.13.

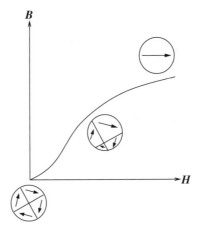

Figure 7.11 Change in domain structure during magnetization of a ferromagnetic material.

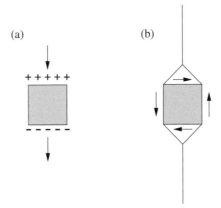

Figure 7.12 (a) Magnetostatic energy around a defect or vacancy enclosed entirely within a domain. (b) The magnetostatic energy can be eliminated if the domain wall intersects the defect and closure domains form.

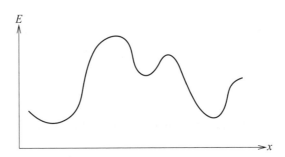

Figure 7.13 Variation of the energy of a Bloch wall with position in an imperfect crystal. The energy minima occur when walls intersect defects or vacancies.

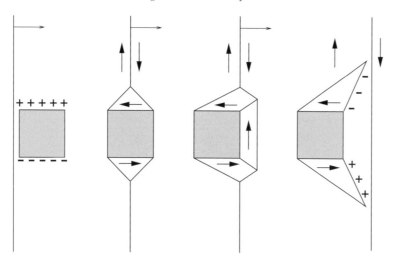

Figure 7.14 Formation of spike domains as a domain boundary moves past a defect.

A schematic of the motion of a boundary past an imperfection is shown in Fig. 7.14. When the boundary moves due to a change in the applied field, the domains of closure cling to the imperfection forming spike-like domains, which continue to stretch as the boundary is forced to move further. Eventually the spike domains snap off and the boundary can move freely again. The field required to snap the spike domains off the imperfections corresponds to the coercive force of the material. A photograph of spike domains in single crystals of silicon iron, highlighted using the colloidal magnetite method, is shown in Fig. 7.15.[28]

When the spikes snap from the domain boundary, the discontinuous jump in the boundary causes a sharp change in flux. The change in flux can be observed by winding a coil around the specimen and connecting it to an amplifier and loudspeaker. Even if the applied field is increased very smoothly, crackling noises are heard from the loudspeaker. This phenomenon is known as the Barkhausen effect. It was first observed in 1919 and provided the first experimental evidence for the existence of domains.[29] Figure 7.16 is a schematic enlargement of a portion of a magnetization curve, showing the sharp changes in magnetization produced by the Barkhausen mechanism.

Eventually the applied field is sufficient to eliminate all domain walls from the sample, leaving a single domain, with its magnetization pointing along the easy axis oriented most closely to the external magnetic field. Further increase in magnetization can only occur by rotating the magnetic dipoles from the easy axis of magnetization into the direction of the applied field. In crystals with large magnetocrystalline anisotropy, large fields can be required to reach the saturation magnetization.

Figure 7.15 Colloidal magnetite pattern of spike domains on single crystals of silicon iron. The lighter-colored regions are the domain boundaries. Width of field 0.4 mm. From Ref. 28. Copyright (1949) by the American Physical Society.

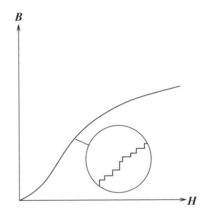

Figure 7.16 Enlargement of the magnetization curve to show the Barkhausen noise.

As soon as the magnetic field is removed, the dipoles rotate back to their easy axis of magnetization, and the net magnetic moment along the field direction decreases. Since the dipole rotation part of the magnetization process did not involve domain wall motion it is entirely reversible. Next, the demagnetizing field in the sample initiates the growth of reverse magnetic domains which allow the sample to be partially demagnetized. However, the domain walls are unable to fully reverse their motion back to their original positions. This is because the demagnetization

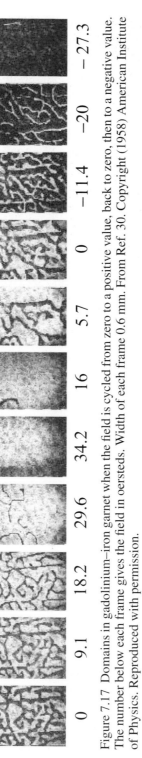

Figure 7.17 Domains in gadolinium–iron garnet when the field is cycled from zero to a positive value, back to zero, then to a negative value. The number below each frame gives the field in oersteds. Width of each frame 0.6 mm. From Ref. 30. Copyright (1958) American Institute of Physics. Reproduced with permission.

process is driven by the demagnetizing field, rather than an applied external field, and the demagnetizing field is not strong enough to overcome the energy barriers encountered when the domain walls intersect crystal imperfections. As a result, the magnetization curve shows hysteresis, and some magnetization remains in the sample even when the field is removed completely. The coercive field is defined to be the additional field, applied in the reverse direction, which is needed to reduce the magnetization to zero.

So we see that the hysteresis properties of a sample depend in large part on its purity and quality. This means that we can engineer materials to optimize their properties for specific applications. For example a sample with many defects or impurities will require a large field to magnetize it, but will retain much of its magnetization when the field is removed. As we mentioned in Chapter 2, materials which are characterized by high remanence and large coercive field are known as *hard* magnetic materials, and are important as permanent magnets. High-purity materials, with few dislocations or dopants, are easily magnetized and demagnetized – these are known as *soft* magnetic materials. Soft magnetic materials are used in electromagnets and transformer cores, where they must be able to reverse their direction of magnetization rapidly.

Finally, in Fig. 7.17, we show some real photographs of the domain structure in gadolinium–iron garnet as the field is cycled from zero to a value large enough to create a single domain oriented in one direction, back to zero, and then to a large value in the opposite direction.[30] The dark and light regions, obtained using the magneto-optic Faraday effect, which we will discuss in Chapter 12, indicate domains of opposite magnetization. The hysteresis can be seen by comparing the third and sixth fames, which occur at similar fields, (the first while the field is increasing, and the second while it is being reduced from its maximum value) but show quite different domain structures.

Homework

Exercises

7.1   Why do domains form in ferromagnetic materials? What are the various contributions to the total energy of a ferromagnetic material, and how do they determine the size and shape of domains?

7.2   Sketch and explain how the domain structure of an initially unmagnetized sample of a ferromagnetic material changes during magnetization to saturation.

7.3   What characteristics would you expect to see in the magnetization curve and hysteresis loop of a perfect (defect-free) ferromagnetic material with a large magnetocrystalline anisotropy? Suggest an application for such a material.

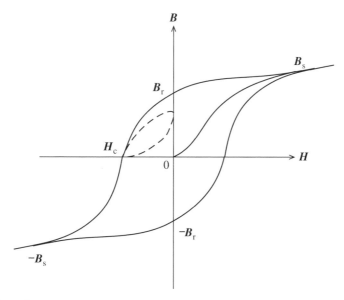

Figure 7.18 Major (solid line) and minor (dashed line) hysteresis loops for a ferromagnetic material.

7.4 What characteristics would you expect in the magnetization curve and hysteresis loop of a ferromagnetic material which has many defects? Suggest an application for such a material.

7.5 Figure 7.18 shows the major hysteresis loop for a ferromagnetic material (solid line) and a minor hysteresis loop (dashed line). We have discussed the domain-based explanation for the form of the major hysteresis loop. Describe the corresponding variation in domain pattern around the *minor* hysteresis loop.

7.6 Figure 7.19 again shows the major hysteresis loop for our ferromagnetic material (solid line), and the dashed line shows a spiral path which returns the material back to the unmagnetized state. Give a domain-based explanation for the form of the path. How else might we convert a ferromagnetic material to an unmagnetized state?

7.7 The boundary between domains is called a domain wall. The exchange energy cost per square meter, σ_{ex}, within a domain wall is given by

$$\sigma_{\mathrm{ex}} = \frac{k_B T_C}{2} \left(\frac{\pi}{N}\right)^2 N \frac{1}{a^2} \ \mathrm{J/m^2}, \tag{7.1}$$

where $N + 1$ is the number of atomic layers in the wall, and a is the spacing between the atoms. The anisotropy energy cost per square meter, σ_A, is given by

$$\sigma_A = K N a \ \mathrm{J/m^2}, \tag{7.2}$$

where K, the magnetocrystalline anisotropy constant, is a measure of the cost of not having all the atoms aligned along easy axes.

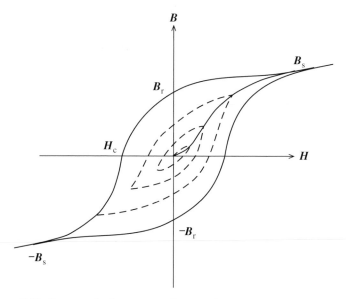

Figure 7.19 Return of a ferromagnetic material to the unmagnetized state.

(a) Plot the form of the exchange energy cost, the anisotropy energy cost and the sum of these two energy costs, for iron, for which $K = 0.5 \times 10^5$ J/m^3, $a = 0.3$ nm, and $T_C = 770\,°$C.

(b) Assuming that the exchange and anisotropy energies are the principal contributors to the domain wall energy, derive an expression for the number of atomic layers in a domain wall, as a function of the Curie temperature, the anisotropy constant, and the atomic spacing.

(c) Calculate the thickness of a domain wall in iron. How much energy is stored in 1 m^2 of an iron domain wall?

Further reading

C. Kittel and J.K. Galt, Ferromagnetic domain theory. *Solid State Physics* **3** 437, 1956.
E.A. Nesbitt, *Ferromagnetic domains*. Bell Telephone Laboratories 1962.
B.D. Cullity, *Introduction to magnetic materials*. Addison-Wesley, 1972, Chapter 9.
D. Jiles, *Introduction to magnetism and magnetic materials*. Chapman & Hall, 1996, Chapters 6 and 7.

8

Antiferromagnetism

"A large number of antiferromagnetic materials is now known; these are generally compounds of the transition metals containing oxygen or sulphur. They are extremely interesting from the theoretical viewpoint but do not seem to have any applications."

Louis Néel, *Magnetism and the local molecular field*, Nobel Lecture, December, 1970.

Now that we have studied the phenomenon of cooperative ordering in ferromagnetic materials, it is time to study the properties of *anti*ferromagnets. In antiferromagnetic materials, the interaction between the magnetic moments tends to align adjacent moments antiparallel to each other. We can think of antiferromagnets as containing two interpenetrating and identical sublattices of magnetic ions, as illustrated in Fig. 8.1. Although one set of magnetic ions is spontaneously magnetized below some critical temperature (called the Néel temperature, T_N) the second set is spontaneously magnetized by the same amount in the opposite direction. As a result, antiferromagnets have no net spontaneous magnetization, and their response to external fields at a fixed temperature is similar to that of paramagnetic materials – the magnetization is linear in the applied field, and the susceptibility is small and positive. The temperature dependence of the susceptibility above the Néel temperature is also similar to that of a paramagnet, but below T_N it decreases with decreasing temperature, as shown in Fig. 8.2.

The first conclusive evidence for the magnetic structure of antiferromagnets was provided by neutron diffraction experiments. We will begin this chapter by reviewing the physics of neutron diffraction, and showing some examples of its successes. Then we will use the localized-moment theory to understand the observed temperature dependence of susceptibility in antiferromagnets. Although, like the paramagnets, antiferromagnetic materials do not find many technological applications, the theoretical analysis is also relevant for the technologically important ferrimagnets which we will discuss in the next chapter. Finally, we will explain the

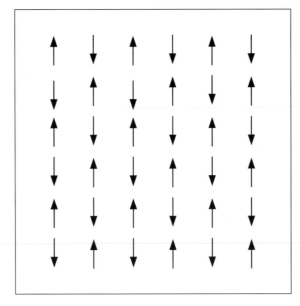

Figure 8.1   Ordering of magnetic ions in an antiferromagnetic lattice.

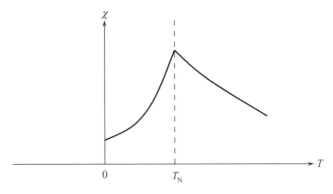

Figure 8.2   Typical temperature dependence of susceptibility in an antiferromagnet.

origin of the antiferromagnetic ordering in some representative magnetic materials by analyzing the nature of the chemical bonding between the magnetic ions.

8.1 Neutron diffraction

The first direct evidence for the existence of antiferromagnetic ordering was provided in 1949, when Shull and Smart[31] obtained the neutron diffraction spectrum of manganese oxide, MnO. Their data showed that the spins on the Mn^{2+} ions are divided into two sets, one antiparallel to the other. Before this breakthrough, the only evidence for antiferromagnetism was the agreement between the observed

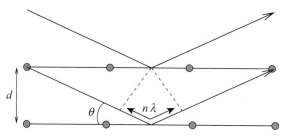

Figure 8.3 Bragg diffraction from planes of atoms. Diffraction peaks are only observed in directions which satisfy the Bragg condition.

temperature dependence of the susceptibility, and the curves predicted using the Curie–Weiss theory. In this section we will review the basics of neutron diffraction and its use in investigating magnetic structure. For an extensive description, see Ref. 32.

Neutron diffraction is able to determine the magnetic ordering of materials, because neutrons have a magnetic moment and so are scattered by the magnetic moments of electrons. This is in contrast to x-rays, which are scattered by electron density and so are not sensitive to magnetic ordering. All diffraction methods are sensitive to the symmetry of the material, and magnetic ordering reduces the symmetry of a material compared to the same material with the magnetic moments oriented randomly. As a result, the neutron diffraction pattern of an antiferromagnet is *different* above and below the Néel temperature.

Just like x-rays, the wavelength, λ, of diffracted neutrons obeys the Bragg equation,

$$n\lambda = 2d \sin \theta. \tag{8.1}$$

The geometry for Bragg diffraction is shown in Fig. 8.3. Each plane of atoms scatters the incident beam in all directions, and most of the scattered beams interfere destructively. Diffraction peaks can only be observed in the directions given by the Bragg equation, where the path difference between scattered beams is a whole number of wavelengths, and constructive interference occurs.

However the number of lines which are actually observed in a diffraction pattern can be fewer than those predicted by the Bragg equation because of the crystal symmetry. This is illustrated for the (100) reflection from a body-centered cubic lattice in Fig. 8.4. Planes (1) and (3) are the (100) planes, and plane (2) is the intermediate plane which contains the body-centered atom. Let's imagine that the crystal is oriented such that the beams scattered from planes (1) and (3) are in phase by the Bragg condition. Then the difference between beams scattered from (1) and (3) must be an integer number of wavelengths, $n\lambda$. It's obvious from the figure that the distance difference between beams scattered from (1) and (2) or (2)

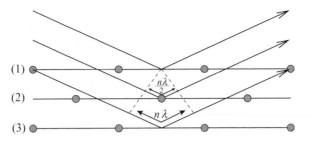

Figure 8.4 Bragg diffraction in a body-centered cubic lattice.

and (3) is exactly *half* of that between those scattered from (1) and (3), that is, a half-integer number of wavelengths. So the reflection from plane (2) is exactly out of phase with that from planes (1) and (3). As a result, the diffracted beams interfere destructively and there is no (100) diffraction line.

If the atoms in plane (2) are different from those in planes (1) and (3), then the beams scattered from (2) will have a different amplitude to those scattered from (1) and (3), and the diffracted beams will no longer cancel. In this case the (100) line will be visible. For neutron scattering, a different orientation of the magnetic moment causes a different scattering amplitude. So if the material orders such that, for example, the atoms in the odd-numbered planes are all up-spin, and those in the even-numbered planes are down-spin, then the (100) line will in fact be present. So as an antiferromagnet is cooled below its Néel temperature, additional lines appear in the neutron diffraction spectrum, indicating the onset of magnetic ordering. These lines are called superlattice lines.

MnO has the face-centered cubic rock-salt structure shown in Fig. 8.5(a). Below the Néel temperature the magnetic moments in each (111) plane align parallel to each other, but they are in opposite directions in successive planes. This magnetic ordering is shown in Fig. 8.5(b). For the face-centered cubic lattice, it turns out that the diffraction line corresponding to the (*hkl*) plane only appears if the Miller indices, *h*, *k* and *l*, are either all odd or all even. The neutron diffraction spectrum of MnO above the Néel temperature is shown in the lower part of Fig. 8.6. As predicted, the (100) and (110) peaks are missing. Below the Néel temperature, the unit cell size doubles, and many more lines appear in the spectrum, as shown in the upper part of Fig. 8.6. Detailed analysis of the spectrum confirms the magnetic ordering shown in Fig. 8.5.

In addition to its sensitivity to magnetic ordering, neutron diffraction has a number of other advantages over more common diffraction techniques such as x-ray diffraction. First, the neutron scattering amplitude varies in an irregular way with atomic number. So neutrons are able to distinguish elements which are adjacent in the periodic table, such as Fe and Co. This is important in the study of ordering in

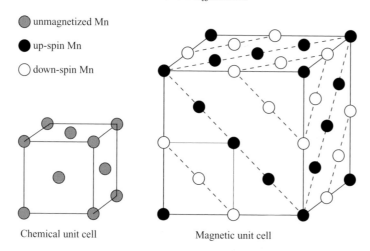

Figure 8.5 Structure of MnO.

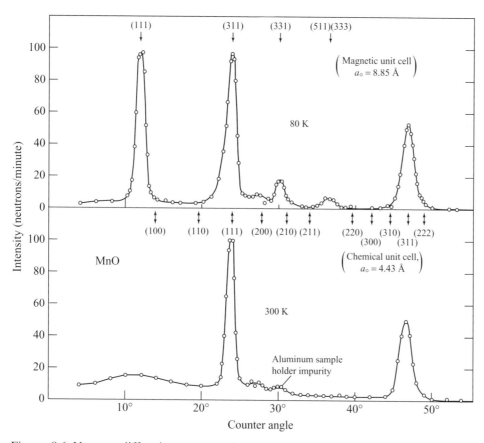

Figure 8.6 Neutron diffraction patterns for MnO at room temperature (lower panel) and at 80 K (upper panel). After Ref. 31. Copyright (1949) by the American Physical Society.

magnetic alloys. (By contrast, the x-ray scattering amplitude is proportional to the atomic number.) In addition, the wavelength of a neutron beam is approximately equal to a typical atomic spacing at room temperature. We can check this using the de Broglie relation, that wavelength is related to the momentum by $\lambda = h/p$, where h is Planck's constant. A neutron has three translational degrees of freedom, so its momentum, p, is determined from $p^2/2m_N = 3k_BT$ where m_N is the mass of a neutron. Combining these two expressions gives a neutron wavelength of 1.49 Å at 20 °C.

8.2 Weiss theory of antiferromagnetism

The Weiss localized moment theory can be applied to antiferromagnets using a similar formalism as for para- and ferromagnets. The algebra was first worked out by Néel,[33] who showed that the observed temperature dependence of the suscepti- bility could be explained by the magnetic ordering which we are now familiar with as antiferromagnetism. In fact the Weiss theory works rather well for antiferromag- nets, since most antiferromagnetic materials are ionic salts with localized magnetic moments.

Before the publication of Néel's classic paper, it was known empirically that the susceptibility of antiferromagnets depends on the temperature as shown in Fig. 8.7. Above the Néel temperature, T_N, the equation of the susceptibility line is

$$\chi = \frac{C}{T - (-\theta)}. \tag{8.2}$$

The susceptibility has a Curie–Weiss dependence on the temperature but with a *negative* value of θ. Remember (from Section 5.2) that $\theta \propto \gamma$, the molecular field constant. So a negative value for θ suggests the existence of a negative Weiss molecular field, which causes the moments to anti-align! The phase transition to the antiferromagnetic state occurs at T_N, and below this temperature the susceptibility decreases slightly with decreasing temperature.

Let's consider the simplest possible example to see how the Weiss localized- moment theory accounts for this behavior. We will divide the lattice into two struc- turally identical sublattices containing atoms labeled A and B, respectively, and assume that the only important interactions are between nearest neighbor A–B pairs of atoms. So we will ignore both A–A and B–B interactions. Then there will be *two* Weiss molecular fields. The field which acts on the A sublattice is pro- portional, but in the opposite direction, to the magnetization of the B sublattice. That is

$$\boldsymbol{H}_w^A = -\gamma \boldsymbol{M}_B. \tag{8.3}$$

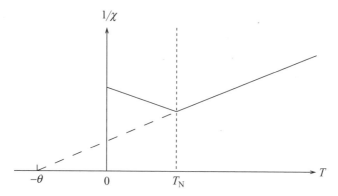

Figure 8.7 Temperature dependence of inverse susceptibility in an antiferromagnet.

Similarly, the field acting on sublattice B is proportional and in the opposite direction to the magnetization of sublattice A:

$$H_{\mathrm{w}}^{B} = -\gamma M_A. \tag{8.4}$$

In both cases the minus signs indicate that the molecular field is opposite to the magnetization of the other sublattice.

8.2.1 Susceptibility above T_N

Above T_N, we can use the Curie law to obtain an expression for the susceptibility, just as we did for non-ideal paramagnets in Section 5.2 and for ferromagnets in Section 6.1.The Curie law tells us that $\chi = M/H = C/T$, so $M = HC/T$. Then, for sublattice A,

$$M_A = \frac{C'(H - \gamma M_B)}{T} \tag{8.5}$$

and for sublattice B,

$$M_B = \frac{C'(H - \gamma M_A)}{T}, \tag{8.6}$$

where H is the external applied field. But the total magnetization, $M = M_A + M_B$, and, solving Eqns 8.5 and 8.6 simultaneously gives

$$M = \frac{2C'H}{T + C'\gamma}. \tag{8.7}$$

So

$$\chi = \frac{M}{H} = \frac{2C'}{T + C'\gamma} \tag{8.8}$$

$$= \frac{C}{T + \theta}. \tag{8.9}$$

This is the Curie–Weiss law with a negative value of θ, as we expected.

8.2.2 Weiss theory at T_N

At the Néel temperature, if there is no external magnetic field, then Eqn 8.5 becomes

$$M_A = \frac{-C'\gamma M_B}{T_N} \tag{8.10}$$

$$= \frac{-\theta M_B}{T_N}. \tag{8.11}$$

But we know that $M_A = -M_B$, therefore

$$\theta = T_N. \tag{8.12}$$

Within the Weiss theory, the Néel temperature is equal to the value of θ obtained from the plot of inverse susceptibility versus temperature. In practice, we find that θ is somewhat larger than T_N. This is not a breakdown of the localized-moment model, but the result of next nearest neighbor interactions which we have not included in our derivation.

8.2.3 Spontaneous magnetization below T_N

Below the Néel temperature, each sublattice is spontaneously magnetized in zero applied field by the molecular field created by the other sublattice. We can write down expressions for the spontaneous magnetization just as we did in Section 6.1.2 for ferromagnets. Again, the most straightforward method of solution is the graphical approach. The spontaneous magnetizations obtained for each sublattice using the graphical method are shown as a function of temperature in Fig. 8.8. At every temperature the net spontaneous magnetization is zero.

8.2.4 Susceptibility below T_N

The susceptibility below T_N depends on the angle between the direction of spontaneous magnetization of the sublattices, and the direction of the applied external field. This is another example of magnetic anisotropy, which we introduced in the

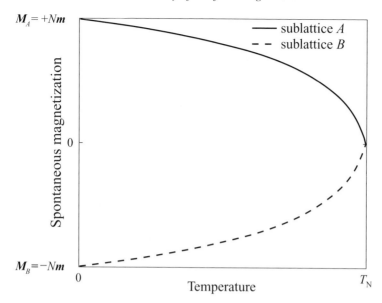

Figure 8.8 Spontaneous magnetization of the A and B sublattices in antiferromagnetic materials below T_N.

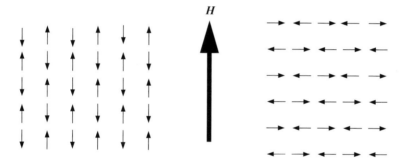

Figure 8.9 Possible orientations of the applied field relative to the magnetization direction in an antiferromagnet.

previous chapter, and which we will discuss in detail in Chapter 10. There are two limiting cases – with the field applied parallel or perpendicular to the magnetization direction, as shown in Fig. 8.9.

Field parallel to magnetization

The spontaneous magnetizations of the A and B sublattices, M_A and M_B, have a Langevin (or Brillouin) function dependence on H and T, as shown in Fig. 8.10. (As before, $\alpha = mH/k_BT$, where m and H represent the magnitudes of the magnetic moment and field vectors respectively). If the external field is applied parallel to the magnetization of the A sublattice, then the magnetization of the A sublattice

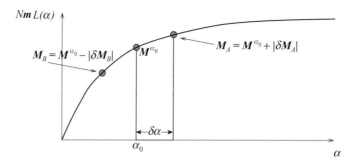

Figure 8.10 Calculation of the susceptibility of an antiferromagnet, with H parallel to M.

increases by an amount δM_A, and that of the B sublattice decreases by δM_B. The material now has a non-zero magnetization, which is

$$M = M_A - M_B \tag{8.13}$$
$$= |\delta M_A| + |\delta M_B|. \tag{8.14}$$

From Fig. 8.10 we can see that, provided the change in magnetization is not too large, the change in magnetization is just the slope of the Brillouin function multiplied by the change in α. But

$$\delta\alpha = \frac{m}{k_{\mathrm{B}}T}\delta H \tag{8.15}$$

$$= \frac{m}{k_{\mathrm{B}}T}(H - \gamma|\delta M_B|). \tag{8.16}$$

Working through the mathematics to calculate the magnetization, then dividing by the external field, gives the following expression for the susceptibility:

$$\chi_\| = \frac{2Nm^2 B'(J,\alpha)}{2k_{\mathrm{B}}T + Nm^2\gamma B'(J,\alpha)}, \tag{8.17}$$

where N is the number of atoms per unit volume and $B'(J,\alpha)$ is the derivative of the Brillouin function with respect to α, evaluated at the point α_0.

The susceptibility tends to zero at zero kelvin, because at zero kelvin the sublattices are perfectly anti-aligned, and there are no thermal fluctuations. Therefore an external field is unable to exert any torque on the magnetic moments. It is interesting to note that a ferromagnetic material below its Curie temperature also follows this expression for the susceptibility. However the change in magnetization as a result of the applied field is negligible compared with the spontaneous magnetization of the ferromagnet and can only be detected at very large external fields. The increase in magnetization of a ferromagnet as a result of a large external field is known as *forced magnetization*.

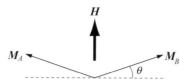

Figure 8.11 Rotation of magnetic moments by a magnetic field applied perpendicular to the direction of magnetization in an antiferromagnet.

Field perpendicular to magnetization

If the external field is applied *perpendicular* to the direction of magnetization, then the atomic magnetic moments are rotated by the applied field, as shown in Fig. 8.11. The rotation creates a magnetization in the field direction, and sets up a molecular field which opposes the magnetization. At equilibrium, the external field, $\boldsymbol{H}$, is exactly balanced by the molecular field, so

$$\boldsymbol{H} = \boldsymbol{H}_{\mathrm{W}} \tag{8.18}$$

$$= 2 \times \boldsymbol{H}_{\mathrm{W}}^{A} \sin\theta \tag{8.19}$$

$$= 2\gamma \boldsymbol{M}_{A} \sin\theta \tag{8.20}$$

$$= \gamma \boldsymbol{M} \tag{8.21}$$

(since $\boldsymbol{M} = 2\boldsymbol{M}_{A} \sin\theta$.) So the susceptibility

$$\chi_{\perp} = \frac{\boldsymbol{M}}{\boldsymbol{H}} = \frac{1}{\gamma}. \tag{8.22}$$

We see that the perpendicular susceptibility, $\chi_{\perp}$, is constant below the Néel temperature.

Powdered samples

In powdered or polycrystalline samples, which have no preferred orientation of the crystals, the susceptibility is obtained by averaging over all possible orientations. Then

$$\chi_{\mathrm{p}} = \chi_{\parallel} \langle \cos^2\theta \rangle + \chi_{\perp} \langle \sin^2\theta \rangle \tag{8.23}$$

$$= \tfrac{1}{3}\chi_{\parallel} + \tfrac{2}{3}\chi_{\perp}. \tag{8.24}$$

The theoretical values of $\chi_{\parallel}$, $\chi_{\perp}$, and χ_{p}, are shown in Fig. 8.12. At all temperatures, $\chi_{\parallel}$ is smaller than $\chi_{\perp}$, and so samples prefer to be oriented with their magnetic moments perpendicular to the applied magnetic field.

In fact, the shape of the χ versus T curve also depends on the magnitude of the applied field. This is another consequence of the magnetic anisotropy; the anisotropy tends to 'pin' the spins along their preferred axis, and a higher field is better able to overcome the pinning.

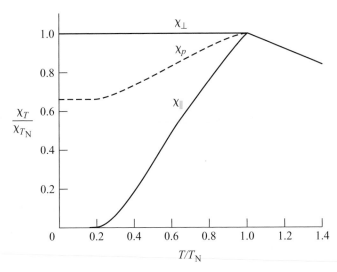

Figure 8.12 Calculated temperature dependence of the susceptibility in antiferromagnetic materials. The curve for $\chi_\parallel$ is calculated for $J = 1$. From Ref. 34. Reprinted by permission of Pearson Education, Inc., Upper Saddle River, NJ.

8.3 What causes the negative molecular field?

In Chapter 6 we saw that the origin of the Weiss molecular field is the quantum mechanical exchange integral, J. A positive exchange integral lowers the energy of parallel spins, relative to those which are aligned antiparallel. We understood this qualitatively by arguing that electrons with the same spin symmetry are prohibited (by the Pauli exclusion principle) from having the same *spatial* symmetry. Therefore they do not occupy the same region of space, and hence they have a lower Coulomb repulsion.

Based on this argument, we might expect that the antiferromagnetic state should always be unstable. For our simple example of the He atom, the exchange integral, J, can never be negative. However in real materials there are usually more than two electrons! The stable state is the one which minimizes the total energy of the system, and can only be predicted if all the many-body interactions are included.

Superexchange

Next we will show how simple valence bonding arguments predict antiferromagnetic ordering in some of the most common antiferromagnets – the magnetic oxides. We will use MnO as our example.

The bonding in MnO is largely ionic, with linear chains of Mn^{2+} and O^{2-} ions running through the crystal. Along each chain direction, the O^{2-} ion has an occupied p orbital oriented along the Mn–O–Mn axis, as shown in Fig. 8.13. Each Mn^{2+} ion contains five 3d electrons, which occupy the 3d orbitals with one electron per orbital and their spins parallel.

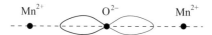

Figure 8.13 Schematic Mn–O–Mn chains in MnO.

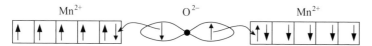

Figure 8.14 Schematic of the superexchange mechanism in MnO.

Next we make the assumption that it is energetically favorable for the valence electrons on the Mn^{2+} and O^{2-} ions to undergo some degree of covalent bonding. Since the O^{2-} ion has a filled shell of electrons, hybridization can only take place by donation of electrons from the O^{2-} ion into the vacant orbitals of the Mn^{2+} ion. Let's assume that our left-most Mn^{2+} ion has up-spin, as shown in Fig. 8.14. Then, since all the Mn orbitals contain an up-spin electron, covalent bonding can only occur if the neighboring oxygen donates its down-spin electron. This leaves an up-spin electron in the oxygen p orbital, which it is able to donate to the next Mn^{2+} ion in the chain. By the same argument, bonding can only occur if the electrons on the next Mn^{2+} ion are down-spin. We see that this oxygen-mediated interaction leads to an overall antiferromagnetic alignment between the Mn^{2+} ions, without requiring negative exchange integrals.

Antiferromagnetism in transition metals

In Chapter 6 we showed that simple band theory arguments explain the presence of ferromagnetism in Fe, Ni and Co, and its absence in Cu and Zn. We only told half the story however. In fact Cr and Mn have complicated antiferromagnetic structures, and to understand this we need to look a little deeper at their electronic structures.

In Chapter 5 we introduced the concept of the Fermi surface – that is, the surface showing the position of the Fermi level, E_F, in k-space. For free electrons the Fermi surface is a sphere, because $E_F = \hbar^2/2m_e k_F^2$. For transition metals, with both d and s bands intersecting the Fermi level, the Fermi surface is much more complicated. As an example, the Fermi surface of chromium, calculated in Ref. 35 using the linear combination of atomic orbitals method, is shown in Fig. 8.15. The figure shows that there are regions of the Fermi surface in which two rather flat surfaces are parallel to each other. When this occurs, an oscillatory spin density develops, with the wavenumber determined by the difference in wavenumber between the two surfaces. If this wavenumber is commensurate with the atomic spacing, we obtain antiferromagnetic ordering. For incommensurate wavenumbers, more complicated spin wave ordering can result.

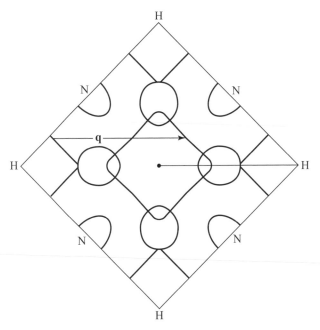

Figure 8.15 An (001) cross-section through the body-centered cubic Brillouin zone showing the Fermi surface of Cr. From Ref. 35. Copyright (1973) by the American Physical Society.

RKKY theory

In rare earth metals, or in alloys of magnetic ions in a non-magnetic metallic host, the magnetic ions are too far apart to interact with each other directly. However a long-range interaction can occur via the non-magnetic conduction electrons. Essentially, a magnetic ion polarizes its surrounding conduction electrons, which, because they are delocalized, transfer their polarization to a second, distant magnetic ion. The resulting interaction between the magnetic ions can be either ferro- or antiferromagnetic, depending on the distance between the ions. The interaction is known as the RKKY interaction (after Ruderman, Kittel, Kasuya and Yosida[36-38]), and was first developed to explain the indirect exchange coupling of nuclear magnetic moments by conduction electrons. The RKKY magnetization of a free electron Fermi gas at zero kelvin around a point magnetic moment is shown in Fig. 8.16.

8.4 Uses of antiferromagnets

Antiferromagnets do not have the wide applicability of ferromagnets because of their lack of spontaneous magnetization. They are, however, closely related structurally to the spontaneously magnetized ferrimagnetic materials which we will

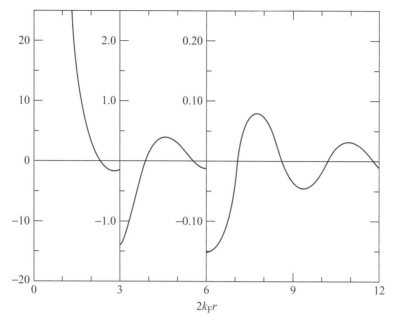

Figure 8.16 Magnetization of free electrons around a point magnetic moment placed at the origin according to RKKY theory. The horizontal axis is $2k_Fr$, where k_F is the Fermi wavevector. The vertical axis is proportional to the magnetization induced by a point source at $r = 0$. From Ref. 16, Kittel, *Introduction to solid state physics*, 7th edn. Copyright (1995) John Wiley & Sons, Inc. Reprinted by permission of John Wiley & Sons, Inc.

study in the next chapter. Therefore they provide a somewhat simpler system in which to test the theoretical models which we will use to explain ferrimagnetism.

One area in which antiferromagnets are starting to find wide applicability is in so-called spin valves (described in Chapter 11) because of a phenomenon called exchange anisotropy or exchange-bias coupling. Exchange anisotropy was first observed over 40 years ago in single-domain particles (100–1000 Å in diameter) of Co (which is ferromagnetic) coated with antiferromagnetic CoO,[39] as shown in Fig. 8.17. Those Co/CoO samples which were cooled in zero field had normal hysteresis behavior, whereas field-cooled samples were observed to have a *shifted* hysteresis loop, as shown schematically in Fig. 8.18. Overall, the coercivity was increased compared with the zero field-cooled sample, and its magnitude was different for increasing and decreasing field.

The observations are explained as follows. The shift occurs during field cooling because the Co ions in the first layer of the CoO experience a ferromagnetic exchange with those in the Co metal and therefore align parallel to them. This alignment persists even when the applied field is removed. Note that the purpose of the field is to give the sample a single alignment direction. In the absence of

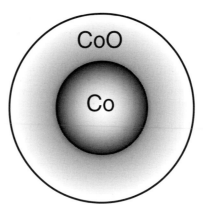

Figure 8.17 Schematic of a core–shell particle consisting of a ferromagnetic Co core, with a surrounding shell of antiferromagnetic CoO.

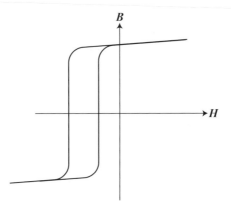

Figure 8.18 Schematic hysteresis loop in a system with exchange anisotropy.

the field, the exchange interaction occurs at all interfaces, resulting in a random distribution of easy directions. The left-most diagram in Fig. 8.19 shows the parallel alignment of the spins on the cobalt ions in the ferromagnetic cobalt layer (all black arrows) with the first row of spins on the cobalt ions in the CoO layer (alternating rows of black arrows and circles representing the oxygen atoms). If the field is subsequently reversed, the spins in the Co metal reverse, but the reversal of spins in the oxide is resisted by the strong magnetocrystalline anisotropy of the CoO. This is represented in the middle diagram of Fig. 8.19. Therefore a large coercive field is needed to reverse the direction of magnetization. Finally, when the field is removed, the Co spins in the CoO at the interface force the Co spins in the Co metal to reverse back into parallel alignment. Therefore the coercive field is reduced or even negative, compared with the non-exchange-biased case.

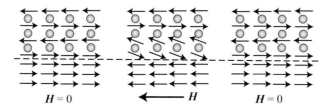

Figure 8.19 Mechanism of hysteresis shift in exchange-biased systems.

In modern spin valve applications, an antiferromagnetic layer is used to pin the direction of magnetization of an adjacent ferromagnetic layer, which is coupled to a second ferromagnetic layer that can rotate in response to an applied field. The resistance of the device is low if both ferromagnetic layers are aligned in the same direction, and high if they are aligned in opposite directions. Therefore the device can be used as a magnetic valve. An excellent review of exchange bias and related effects, including a compilation of materials, experimental techniques for studying them, potential applications and theoretical models, is given in Ref. 40.

Additional applications for antiferromagnetic materials might emerge from the huge current research effort aimed at exploiting materials which show a phase transition from an antiferromagnetic to a ferromagnetic state, with corresponding changes in structural and magnetic properties. Examples of such materials include the so-called colossal magnetoresistive (or CMR) materials. The CMR materials are perovskite-structure manganites in which the ferromagnetic to antiferromagnetic transition is accompanied by a metal–insulator transition. As a result, they show a large change in conductivity when a magnetic field is applied, leading to potential applications as magnetic field sensors. We will discuss CMR materials and other magnetoresistors in Chapter 11 of this book.

Homework

Exercises

8.1   Show that Eqn 8.17, describing the susceptibility when the field is applied parallel to the magnetization direction, reduces to the Curie–Weiss expression (Eqn 8.9) at high temperatures, and to zero at zero kelvin.

8.2   Consider an antiferromagnetic material which has a susceptibility χ_0 at its Néel temperature, T_N. Assuming that the exchange interactions between nearest neighbor A and B ions are much larger than those between A–A and B–B pairs, calculate the values of the susceptibilities which would be measured under the application of fields perpendicular to the magnetization direction at $T = 0$, $T = T_N/2$ and $T = 2T_N$.

To think about

We've seen that the superexchange mechanism leads to antiferromagnetism. Do you think it is likely that ferromagnetic oxides exist? Think about what might happen if you had a Mn^{3+} ion (with four 3d electrons) separated from a Mn^{4+} ion (with three 3d electrons) by an oxygen ion. More about this in Section 11.4.4.

Further reading

B.D. Cullity, *Introduction to magnetic materials*. Addison-Wesley, 1972, Chapter 5.

9

Ferrimagnetism

"To interpret the magnetic properties, I assumed that the predominant magnetic interactions were exerted between the ions placed at sites A and ions placed at sites B, and that they were essentially negative."
Louis Néel, *Magnetism and the local molecular field*, Nobel Lecture, December, 1970.

Finally we have reached the last chapter in our survey of the most important types of magnetic materials. In this chapter we will discuss ferrimagnets. Ferrimagnets behave similarly to ferromagnets, in that they exhibit a spontaneous magnetization below some critical temperature, T_C, even in the absence of an applied field. However, as we see in Fig. 9.1, the form of a typical ferrimagnetic magnetization curve is distinctly different from the ferromagnetic curve.

In fact ferrimagnets are also related to *anti*ferromagnets, in that the exchange coupling between adjacent magnetic ions leads to antiparallel alignment of the localized moments. The overall magnetization occurs because the magnetization of one sublattice is greater than that of the oppositely oriented sublattice. A schematic of the ordering of magnetic moments in a ferrimagnet is shown in Fig. 9.2. We will see in the next section that the observed susceptibility and magnetization of ferrimagnets can be reproduced using the Weiss molecular field theory. In fact the localized moment model applies rather well to ferrimagnetic materials, since most are ionic solids with largely localized electrons.

The fact that ferrimagnets are ionic solids means that they are electrically insulating, whereas most ferromagnets are metals. This results in a wide range of important applications for ferrimagnets, in situations requiring magnetic insulators. In Sections 9.2 and 9.3 we will review the properties of some of the most technologically relevant ferrimagnetic materials – the ferrites and the garnets. At the end of this chapter, we will discuss, just for fun, a new class of materials which has been predicted theoretically, but has not yet been synthesized; the so-called

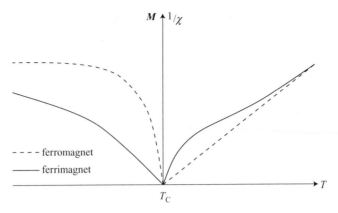

Figure 9.1 Comparison of magnetization and inverse susceptibility in typical ferri- and ferromagnets.

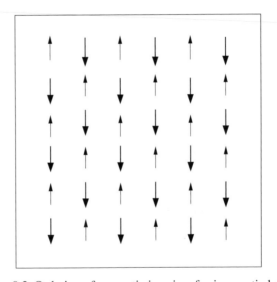

Figure 9.2 Ordering of magnetic ions in a ferrimagnetic lattice.

half-metallic antiferromagnets, which are actually unusual ferrimagnets with zero net magnetization.

9.1 Weiss theory of ferrimagnetism

Néel developed the theory of ferrimagnetism in the same classic paper which contained his theory of antiferromagnetism.[33] The localized-moment picture for ferrimagnets is slightly more complicated than that for antiferromagnets. This time the *A* and *B* sublattices are not structurally identical, and we have to consider at least *three* interactions to reproduce the experimentally observed behavior. These are the interactions between nearest neighbor A–B pairs which tend to align the moments in the two sublattices antiparallel, plus the A–A and B–B interactions. Here A and B can represent either different atomic species, or the same ion types on sites of different symmetry.

The simplest possible model which allows us to reproduce the features of ferrimagnetism must include interactions between A–A, B–B, and A–B ion pairs. We will assume that the A–B interaction drives the antiparallel alignment, and that both A–A and B–B interactions are ferromagnetic. In the following derivation, n is the number of magnetic ions per unit volume, α is the fraction of A ions, β is the fraction of B ions ($= 1 - \alpha$), μ_A is the average magnetic moment of an A ion in the direction of the field at some temperature T, and μ_B is the average moment of a B ion.

Then the magnetization of the A sublattice,

$$\boldsymbol{M}_A = \alpha n \mu_A, \tag{9.1}$$

and the magnetization of the B sublattice,

$$\boldsymbol{M}_B = \beta n \mu_B. \tag{9.2}$$

So the total magnetization,

$$\boldsymbol{M} = \boldsymbol{M}_A + \boldsymbol{M}_B = \alpha n \mu_A + \beta n \mu_B. \tag{9.3}$$

Again there are two Weiss molecular fields, one acting on each of the A and B sublattices, but they are no longer equal in magnitude. The molecular field on the A sublattice is

$$\boldsymbol{H}_{\mathrm{W}}^{A} = -\gamma_{AB} \boldsymbol{M}_B + \gamma_{AA} \boldsymbol{M}_A. \tag{9.4}$$

Similarly, the field acting on sublattice B is given by

$$\boldsymbol{H}_{\mathrm{W}}^{B} = -\gamma_{AB} \boldsymbol{M}_A + \gamma_{BB} \boldsymbol{M}_B. \tag{9.5}$$

The minus signs indicate a contribution to the molecular field which is opposite in direction to the corresponding magnetization.

9.1.1 Weiss theory above T_C

To apply the Weiss theory above the Curie temperature, we assume Curie-law behavior for each sublattice. (This method should now be very familiar!) That is $\chi = M/H_{\mathrm{tot}} = C/T$, so $M = H_{\mathrm{tot}} C/T$, where H_{tot} is the total field, which is the sum of the applied field and the Weiss field. Then, for sublattice A,

$$M_A = \frac{C\left(H + H_W^A\right)}{T} \tag{9.6}$$

and for sublattice B,

$$M_B = \frac{C\left(H + H_W^B\right)}{T}. \tag{9.7}$$

Here H is the external applied field.

Solving for $M = M_A + M_B$, and dividing by the field to obtain the susceptibility, gives

$$\frac{1}{\chi} = \frac{T + C/\chi_0}{C} - \frac{b}{T - \theta}. \tag{9.8}$$

Here

$$\frac{1}{\chi_0} = \gamma_{AB}\left(2\alpha\beta - \frac{\gamma_{AA}}{\gamma_{AB}}\alpha^2 - \frac{\gamma_{BB}}{\gamma_{AB}}\beta^2\right), \tag{9.9}$$

$$b = \gamma_{AB}^2 C\alpha\beta\left[\alpha\left(1 + \frac{\gamma_{AA}}{\gamma_{AB}}\right) - \beta\left(1 + \frac{\gamma_{BB}}{\gamma_{AB}}\right)\right]^2, \tag{9.10}$$

and

$$\theta = \gamma_{AB}C\alpha\beta\left(2 + \frac{\gamma_{AA}}{\gamma_{AB}} + \frac{\gamma_{BB}}{\gamma_{AB}}\right). \tag{9.11}$$

The curve described by Eqn 9.8 is plotted in Fig. 9.3. It is a hyperbola, and intersects the temperature axis at the so-called paramagnetic Curie point, θ_p. At high temperatures the second term in the expression for $\frac{1}{\chi}$ becomes small, and Eqn 9.8 reduces to a Curie–Weiss law:

$$\chi = \frac{C}{T + (C/\chi_0)}. \tag{9.12}$$

This Curie–Weiss equation is plotted as the dashed line in Fig. 9.3.

The Curie–Weiss prediction gives good agreement with experiment, except in the immediate vicinity of the Curie point. Figure 9.4 shows the measured reciprocal susceptibility of magnesium ferrite from Ref. 41, compared with the theoretical prediction, using values of the constants given by Néel.[33] The intersection of the experimental curve with the temperature axis is called the ferromagnetic Curie temperature, θ_f. This is the temperature at which the susceptibility diverges and spontaneous magnetization appears. The experimental θ_f differs slightly from the predicted θ_p, because of short-range magnetic order which persists for a few degrees even above T_C.

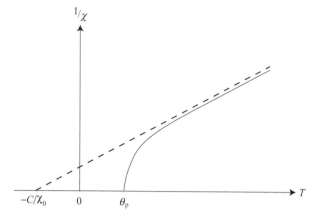

Figure 9.3 Inverse susceptibility as a function of temperature in ferrimagnetic materials, calculated using the Weiss theory. From Ref. 34. Reprinted by permission of Pearson Education, Inc., Upper Saddle River, NJ.

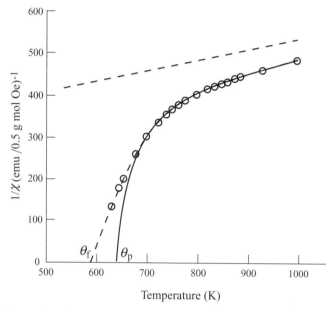

Figure 9.4 Measured and predicted inverse susceptibility of Mg ferrite. From Ref. 34. Reprinted by permission of Pearson Education, Inc., Upper Saddle River, NJ.

9.1.2 Weiss theory below T_C

Below the Curie temperature, each sublattice is spontaneously magnetized, and there is a net observable magnetization,

$$M = |M_A| - |M_B|. \tag{9.13}$$

Each sublattice follows the familiar Brillouin function magnetization curve, so (just as for ferromagnets),

$$M_A = N m_A B \left(J, \frac{m_A H_W^A}{k_B T} \right) \tag{9.14}$$

and

$$M_B = N m_B B \left(J, \frac{m_B H_W^B}{k_B T} \right). \tag{9.15}$$

Here m_A and m_B are the magnetic moments along the field direction on the A and B ions respectively. Substituting for H_W^A and H_W^A,

$$M_A = N m_A B \left(J, \frac{m_A [\gamma_{AA} M_A - \gamma_{AB} M_B]}{k_B T} \right) \tag{9.16}$$

and

$$M_B = N m_B B \left(J, \frac{m_B [\gamma_{BB} M_B - \gamma_{AB} M_A]}{k_B T} \right). \tag{9.17}$$

These equations are not independent – the magnetization of the A sublattice depends on the magnetization of the B sublattice, and vice versa. Therefore the simple graphical method of solution which we used for antiferromagnetic materials cannot be used here, and the equations must be solved numerically.

The resulting spontaneous magnetization curves for typical values of γ_{AB}, γ_{AA} and γ_{BB} are shown in Fig. 9.5. Note that both sublattices *must* have the same Curie point, otherwise at some temperature one of the lattices would have zero moment, and so would not be able to align the moments on the other sublattice.

Because the shapes of the spontaneous magnetization curves for each sublattice depend on the values of all the molecular field constants, and on the distribution of

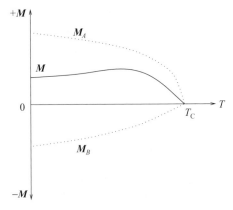

Figure 9.5 Schematic of the spontaneous magnetization of the A and B sublattices (dotted curves), and resultant magnetization (solid curve), in a typical ferrimagnetic material.

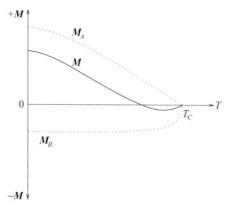

Figure 9.6 Unusual magnetization curves in ferrimagnets.

A- and B-type ions, the net magnetization does not necessarily vary monotonically with temperature. We've already seen one example in Fig. 9.5, where the magnitude of the A sublattice magnetization decreases less rapidly with increasing temperature than the magnitude of the B sublattice magnetization. As a result, the net spontaneous magnetization *increases* with increasing temperature, and goes through a maximum before falling to zero at T_C. Such behavior is displayed, for example, by the cubic spinel $NiO \cdot Cr_2O_3$. Figure 9.6 shows a different case – here the spontaneous magnetization decreases to zero before the Curie temperature is reached, and then the material develops a spontaneous magnetization in the opposite direction. At one temperature, called the *compensation point*, the magnetizations of the two sublattices are exactly balanced and the net magnetization is zero. The compound $Li_{0.5}Fe_{1.25}Cr_{1.25}O_4$ is an example of a material which shows this behavior.

If a material contains more than two sublattices then an even more complicated temperature dependence of the magnetization, including more than one compensation point, can occur. One example which has been synthesized recently[42, 43] is $(Ni_{0.22}Mn_{0.60}Fe_{0.18})_3[Cr(CN)_6]$, a Prussian-blue structure phase in which the transition metal cations form a face-centered cubic array linked by cyanide anions. This material is ferrimagnetic with a Curie temperature of 63 K, and exhibits two magnetization reversals, at 53 and 35 K, as shown in Fig. 9.7. The properties are well described by a three-component Weiss molecular field theory.

9.2 Ferrites

The most technologically important ferrimagnets are the materials known as ferrites. Ferrites are ferrimagnetic oxides, and therefore are electrically insulating. As a result they find applications in situations where the electrical conductivity shown by most ferromagnetic materials would be detrimental. For example they are widely

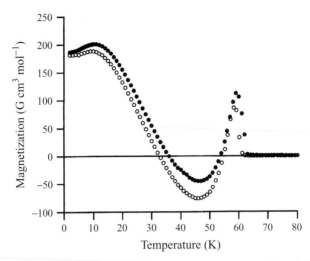

Figure 9.7 Experimental magnetization curves for $(Ni_{0.22}Mn_{0.60}Fe_{0.18})_3[Cr(CN)_6]$. The filled circles show the field-cooled magnetization obtained with decreasing temperature in an external magnetic field of 10 G; the open circles show the remanent magnetization obtained with increasing temperature after the temperature was first lowered in the applied magnetic field of 10 G. From Ref. 43. Copyright (1999) by the American Physical Society.

used in high-frequency applications, because an ac field does not induce undesirable eddy currents in an insulating material.

Ferrites are usually manufactured using ceramic processing techniques. For example, to produce $NiO \cdot Fe_2O_3$, powdered NiO and Fe_2O_3 are mixed together, pressed into shape, and heated. This method has the advantage of allowing easy control of the shape of the magnet by the choice of the mold.

There are two common types of ferrites with different structural symmetries – the cubic ferrites and the hexagonal ferrites.

9.2.1 The cubic ferrites

The cubic ferrites have the general formula $MO \cdot Fe_2O_3$, where M is a divalent ion, such as Mn^{2+}, Ni^{2+}, Fe^{2+}, Co^{2+}, or Mg^{2+}. The earliest technologically useful magnetic material, magnetite, is a cubic ferrite. Magnetite has the formula $FeO \cdot Fe_2O_3$, and is the magnetic mineral contained in lodestone, from which the first compasses for navigation were made.

Cubic ferrites crystallize in the spinel structure (named after the mineral spinel, $MgO \cdot Al_2O_3$). The oxygen anions are packed in a face-centered cubic arrangement such that there are two kinds of spaces between the anions – tetrahedrally coordinated (or A) sites, and octahedrally coordinated B sites. The cations occupy the

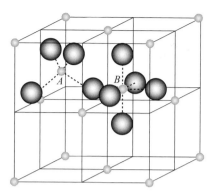

Figure 9.8 Schematic of the spinel structure, showing octahedral and tetrahedral sites occupied by A and B cations.

spaces, although not all of the spaces are occupied. A picture of the spinel structure is shown in Fig. 9.8.

In the *normal* spinel structure, the divalent M^{2+} ions are all on A sites, and the Fe^{3+} ions occupy B sites. Examples of such ferrites include $ZnO \cdot Fe_2O_3$ and $CdO \cdot Fe_2O_3$, both of which are paramagnetic. In the *inverse* spinels, the divalent ions occupy some of the B sites, and the Fe^{3+} ions are divided equally between A and B sites. Examples include Fe-, Co- and Ni ferrite, all of which are ferrimagnetic. The spin moments of all the Fe^{3+} ions on the octahedral sites are aligned parallel to one another, but directed oppositely to the spin moments of the Fe^{3+} ions occupying the tetrahedral positions. Therefore the magnetic moments of all Fe^{3+} ions cancel and make no net contribution to the magnetization of the solid. However all the divalent ions have their moments aligned parallel to one another, and it is this total moment which is responsible for the net magnetization. Thus the saturation magnetization of a ferrimagnetic solid can be calculated from the product of the net spin magnetic moment of each divalent cation and the concentration of divalent cations.

Magnetization curves for a range of cubic ferrites are shown in Fig. 9.9. It is clear that the saturation magnetization and the Curie temperature vary markedly between different compounds. In addition, solid solutions of mixed ferrites can be formed readily, allowing the values of these properties to be tuned precisely for specific applications.

The cubic ferrites are soft, and so are easily magnetized and demagnetized. Combined with their high permeability and saturation magnetization, and low electrical conductivity, this makes them particularly appropriate as cores for induction coils operating at high frequencies. Their high permeability concentrates flux density inside the coil and enhances the inductance, and their high electrical resistivity reduces the formation of undesirable eddy currents.

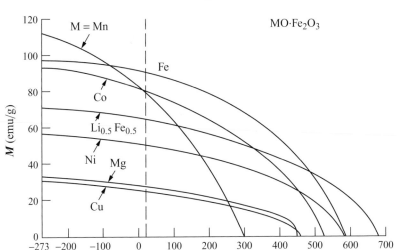

Figure 9.9 Magnetization curves of some cubic ferrites. From Ref. 34. Reprinted by permission of Pearson Education, Inc., Upper Saddle River, NJ.

A history lesson – ferrite core memories

Before the widespread adoption of transistor-based random access memory in computers, memories composed of ferrite cores connected by a network of wires were used. The production of ferrite cores was an important industry – in 1968 alone, more than 15 billion were produced. A schematic of such a ferrite core memory is shown in Fig. 9.10, where the gray rectangular blobs are the ferrite cores, and the black lines are wires connecting them. Each core can be used to store a single bit of information, because it has two stable magnetic states, corresponding to opposite alignments of the remanent flux densities. To switch a core at a particular intersection requires the coincidence of two currents, either of which must be insufficient to exceed the threshold of the core's hysteresis loop on its own.

The most important feature of ferrites which made them suitable for memory applications is their square-shaped hysteresis loops. The origin of the square shape is the large magnetocrystalline anisotropy, which we will discuss in detail in the next chapter. A typical ferrite hysteresis loop is shown in Fig. 9.11. The advantages of the square hysteresis loop are that the remanent magnetization is close to the saturation magnetization, and that a well-defined applied field slightly greater than the coercive field will switch the magnetization direction.

Other desirable characteristics are fast switching times τ, minimal temperature variation (and therefore a high T_C), mechanical strength (allowing small cores to be produced), and low magnetostriction. A widely used material was $Mg_{0.45}Mn^{2+}_{0.55}Mn^{3+}_{0.23}Fe_{1.77}O_4$, which has the characteristics shown in Table 9.1.

Table 9.1 *Important characteristics of*
$Mg_{0.45}Mn^{2+}_{0.55}Mn^{3+}_{0.23}Fe_{1.77}O_4.$

coercivity,	H_c	72 A/m
residual induction,	B_r	0.22 Wb/m^2
saturation induction,	B_s	0.36 Wb/m^2
Curie temperature,	T_C	300 °C
switching time,	τ	0.005 μs A/m

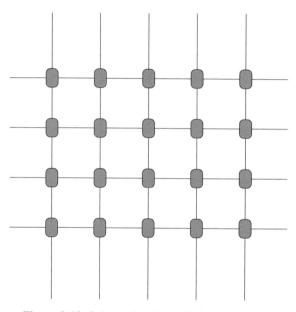

Figure 9.10 Schematic of a ferrite core memory.

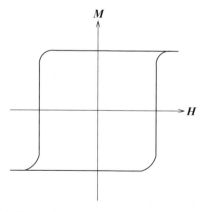

Figure 9.11 Square hysteresis loop typical of cubic ferrites.

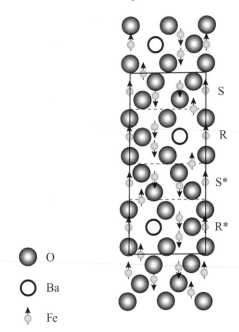

Figure 9.12 Structure of barium ferrite.

9.2.2 The hexagonal ferrites

The most important of the hexagonal ferrites is barium ferrite, $BaO \cdot 6Fe_2O_3$. Barium ferrite crystallizes in the hexagonal magnetoplumbite structure (Fig. 9.12). The magnetoplumbite structure contains 10 oxygen layers in its elementary unit cell, and is constructed from four building blocks, labeled S, S*, R and R* in the figure. The S and S* blocks are spinels with two oxygen layers and six Fe^{3+} ions. Four of the Fe^{3+} ions are in octahedral sites and have their spins aligned parallel to each other (say up-spin), and the other two are in tetrahedral sites, with the opposite spin direction to the octahedral iron ions. The S and S* blocks are equivalent but rotated 180° with respect to each other. The R and R* blocks consist of three oxygen layers, with one of the oxygen anions in the middle layer replaced by a barium ion. Each R block contains six Fe^{3+} ions, five of which are in octahedral sites with three up-spin and two down-spin, and one of which is coordinated by five O^{2-} anions and has up-spin. The net magnetic moment per unit cell is 20 μ_B.

Hexagonal ferrites are used widely as permanent magnets. They are magnetically hard (unlike the cubic ferrites which are magnetically soft) with typical coercivities of around 200 kA/m. Also, they are cheap to produce by ceramic processing methods, and can be powdered and formed easily into any required shape.

9.3 The garnets

The garnets have the chemical formula $3M_2O_3 \cdot 5Fe_2O_3$, where M is yttrium or one of the smaller rare earths towards the right hand side of the lanthanide series (Gd to Lu). All cations in garnets are trivalent (in contrast to the ferrites, which contain some divalent and some trivalent cations). As a result there is no possibility of electron hopping through the material, and the resistivity of garnets is extremely high. Therefore they are used in very high frequency (microwave) applications, where even the ferrites would be too conductive.

The garnets are rather weakly ferrimagnetic. As an example, in yttrium–iron garnet, the yttrium does not have a magnetic moment (since it does not have any f electrons), so the net moment is due entirely to the unequal distribution of Fe^{3+} ions in up- and down-spin sites. The antiferromagnetic superexchange interaction results in three up-spin electrons for every two down-spin electrons, and a net magnetic moment of 5 μ_B per formula unit. Since the formula unit is very large, this leads to a small magnetization per unit volume. In the rare earth garnets, the magnetic moment of the R^{3+} ion also contributes, and this leads to a compensation point in the magnetization curve.

Since the rare earths readily substitute for one another, and Fe^{3+} can be easily replaced by Al^{3+} or Ga^{3+}, it is possible to tune the compensation point, saturation magnetization, anisotropy and lattice constant for specific applications.

9.4 Half-metallic antiferromagnets

Half-metallic antiferromagnets are a class of materials which have been predicted theoretically[44, 45] but not yet synthesized. We include them here in part for some light entertainment, but also to illustrate that there is still great potential in the search for new magnetic materials with novel and possibly technologically relevant properties.

Half-metallic materials are defined to be insulating for one spin direction (down-spin say) but metallic for the other spin channel (up-spin). As such, the Fermi energy is in the band gap for the down-spin electrons, but is in a region of finite density of states for up-spin electrons. A consequence of the half-metallicity is that the spin magnetization is always an integer number of Bohr magnetons per unit cell. In a half-metallic antiferromagnet this integer is zero, so that there is no net magnetization. Half-metallic antiferromagnets are really *ferrimagnets* in which the magnetizations of the two different sublattices exactly cancel out.

The properties of half-metallic antiferromagnets are unusual. First, they are non-magnetic metals in which electric current, carried by electrons near the Fermi

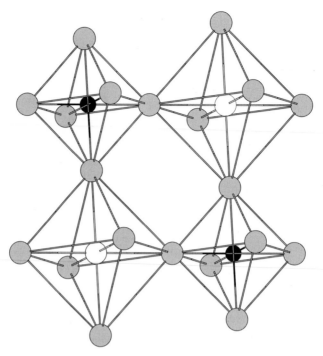

Figure 9.13 Schematic of the double perovskite crystal structure. The black and white spheres are transition metal ions (Mn^{3+} and V^{3+} in our example), surrounded by octahedra of gray oxygen anions, and the La^{3+} cations (not shown) lie between the octahedra. The octahedra around different transition metal cations can be different sizes. From Ref. 45. Copyright (1998) by the American Physical Society.

level, is fully spin-polarized! However since there is no net internal magnetization, half-metallic antiferromagnets do not generate a magnetic field, in spite of their fully magnetized currents. This is a particularly desirable property for example in spin-polarized scanning tunneling microscopy, which allows one to obtain an atomic scale map of spin-resolved information. Currently such experiments are complicated by the existence of a permanent magnetic tip (required to produce the spin-polarized electrons) close to the magnetic surface being investigated. Also a novel form of superconductivity has been proposed.

The most promising candidate materials for half-metallic antiferromagnetism have the double perovskite structure shown in Fig. 9.13. An example which has been shown computationally to have the required band structure is La_2VMnO_6. Here the Mn^{3+} ion has a low spin d^4 configuration, resulting in a net magnetic moment of 2 μ_B, and the V^{3+} ion is d^2 and so also has a moment of 2 μ_B. The most stable state is predicted to have the V^{3+} and Mn^{3+} sublattices aligned antiferromagnetically.

Homework

Exercises

9.1 Review question 1

(a) Outline the major similarities and differences between ferromagnetic and ferrimagnetic materials.

(b) The approximate values of spontaneous magnetization M (normalized by dividing by the saturation magnetization, M_s) for magnetite, Fe_3O_4, as a function of T/T_C were measured by Weiss[18] and are given in the table below:

M/M_s	0.92	0.88	0.83	0.77	0.68	0.58	0.43	0.32	0.22	0.03
T/T_C	0.23	0.33	0.43	0.54	0.66	0.78	0.89	0.94	0.95	0.98

Plot these values, and compare your graph with the curve derived from the Langevin–Weiss theory of ferromagnetism shown in Fig. 6.3. Comment!

(c) Calculate the saturation magnetization for magnetite (Fe_3O_4), given that each cubic unit cell contains 8 Fe^{2+} and 16 Fe^{3+} ions, and that the unit cell edge length is 0.839 nm. For the ferrites, it's safe to assume that the orbital angular momentum is quenched. Also, remember that the magnetization will be measured along the direction of applied field, so when you calculate the magnetic moment per atom, it's the magnetic moment along the field direction that you are interested in.

(d) Design a cubic mixed-ferrite material that has a saturation magnetization of $5.25 \times 10^5 A/m$. (Assume that substituting an iron ion with another ion from the first row transition elements does not change the lattice constant significantly.) What is the saturation flux density of your material? Give your answer (i) in SI units and (ii) in cgs units.

(e) Outline the major similarities and differences between antiferromagnetic and ferrimagnetic materials.

(f) Explain how the superexchange interaction leads to antiferromagnetic coupling between the magnetic ions in ferrimagnetic materials. How would you expect the strength of the superexchange interaction to vary if the cation–oxygen–cation bond angle were increased or decreased from $180°$?

9.2 Review question 2

Cubic nickel ferrite has the chemical formula $NiO·Fe_2O_3$. The structure consists of close-packed planes of oxygen anions, with the nickel ions occupying tetrahedral sites, and the iron ions evenly distributed between octahedral and tetrahedral sites. Each unit cell contains eight formula units.

(a) What are the charges and electronic structures of the nickel and iron ions?

(b) Cations occupying tetrahedral sites have the opposite spin direction to cations occupying octahedral sites. Explain in a few words why this occurs. What is the name of the theory that you have described? As a result of this ferrimagnetic ordering, what net magnetic moment do the iron ions contribute?

(c) The unit cell edge length of nickel ferrite is 8.34 Å . What is the saturation magnetization of nickel ferrite?

(d) Hall effect measurements on metallic, elemental nickel indicate that the number of free electrons per atom of Ni is 0.54. Based on your result, how many d electrons per atom are there in metallic Ni? (HINT – remember that only the s electrons are free and contribute to the conductivity. All the remaining valence electrons must therefore be d electrons).

(e) In ferromagnetic metals, the d-electron band splits into a lower-energy band which is occupied by the up-spin electrons, and a higher-energy band for the down-spin electrons. Only the d electrons contribute to the magnetic moment, and the magnitude of the magnetic moment is determined by the difference between the numbers of up- and down-spin electrons. In Ni, all five of the up-spin d bands are filled. (i) How many down-spin d bands are filled? Sketch the density of states of ferromagnetic Ni. (ii) What is the magnitude of the magnetic moment per atom of Ni?

(f) Elemental Ni crystallizes in the fcc structure with a cubic unit cell edge length of 3.52 Å . How many atoms are there per unit cell? What is the magnetic moment per unit cell? What is the saturation magnetization of elemental Ni?

(g) Compare your calculated saturation magnetizations for nickel ferrite and nickel. Comment on possible applications for both materials.

10

Anisotropy

"... could it work so much upon your shape
As it hath much prevail'd on your condition,
I should not know you, Brutus."
William Shakespeare, *Julius Caesar*, The Oxford Shakespeare, 1914.

The term *magnetic anisotropy* refers to the dependence of the magnetic properties on the direction in which they are measured. The magnitude and type of magnetic anisotropy affect properties such as magnetization and hysteresis curves in magnetic materials. As a result the nature of the magnetic anisotropy is an important factor in determining the suitability of a magnetic material for a particular application. The anisotropy can be either intrinsic to the material, as a result of its crystal chemistry or its shape, or it can be induced by careful choice of processing method. In this chapter we will discuss both intrinsic and induced anisotropies in some detail.

10.1 Magnetocrystalline anisotropy

In Chapter 7 we introduced the concept of magnetocrystalline anisotropy – that is, the tendency of the magnetization to align itself along a preferred crystallographic direction. We also defined the magnetocrystalline anisotropy energy to be the energy difference between samples magnetized along easy and hard directions. The magnetocrystalline anisotropy energy can be observed by cutting a $\{110\}$ disk from a single crystal of material as shown in Fig. 10.1, and measuring the **M–H** curves along the three high-symmetry crystallographic directions ([110], [111] and [001]) contained within the disk.

Schematic results for single-crystal samples of ferromagnetic metals such as iron and nickel were shown in Fig. 7.4. Body-centered cubic Fe has the $\langle 100 \rangle$ direction as its easy axis. In Ni, which is face-centered cubic, the easy axis is $\langle 111 \rangle$. Note that the final value of the spontaneous magnetization is the same, no matter which

123

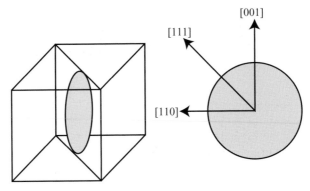

Figure 10.1 Preparation of a sample for measuring the magnetocrystalline anisotropy energy.

axis the field is applied along, but the field required to reach that value is distinctly different in each case.

10.1.1 Origin of magnetocrystalline anisotropy

The energy required to rotate the spin system of a domain away from the easy direction is actually just the energy required to overcome the spin–orbit coupling. When an applied field tries to reorient the direction of the electron spin, the orbit also needs to be reoriented, because of the spin–orbit coupling. However the orbit is in general also strongly coupled to the lattice, and so the attempt to rotate the spin axis is resisted. This is illustrated schematically in Fig. 10.2. Part (a) shows the magnetic moments aligned along the easy axis (say) with the orbital components (which are not spherical because of the spin–orbit coupling) aligned with their long axes along the x axis. For this particular crystal this is a favorable energy configuration. Part (b) shows the result of forcing the magnetic spins to align along the x axis by applying an external magnetic field. The orbital components no longer have favorable overlap with each other or with the lattice.

In most materials the spin–orbit coupling is fairly weak, and so the magnetocrystalline anisotropy is not particularly strong. In rare earth materials, however, the spin–orbit coupling is strong because rare earth elements are heavy. Once magnetized, a large field must be applied in the direction opposite to the magnetization in order to overcome the anisotropy and reverse the magnetization. Therefore rare earth materials are often used in applications such as permanent magnets, where a large coercive field is required.

Schematic magnetization curves for terbium, Tb, which is hexagonal, with the easy magnetization axis in the c plane, are shown in Fig. 10.3. When the field is applied perpendicular to the easy axis, only around 80% of the spontaneous

(a)

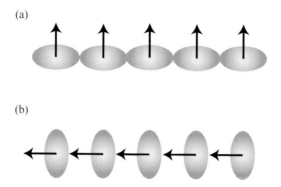

(b)

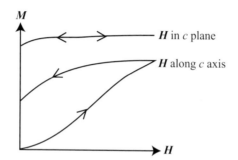

Figure 10.2 Interaction between spin and orbit degrees of freedom.

Figure 10.3 Schematic magnetization curves for Tb, with the field applied along and perpendicular to the easy axis.

magnetization is obtained, even at fields as large as 400 kOe. This is because the strong magnetocrystalline anisotropy resists rotation of the magnetization out of the easy axis. Hysteresis is observed, in spite of the magnetization occurring by what should be reversible rotation of the magnetization, because the strong spin–orbit coupling also leads to a large magnetostriction. This magnetostriction forms mechanical twins along the direction of magnetization, and the twin boundaries must be reoriented before the magnetization can relax.

Terbium has 8 electrons in its unfilled 4f shell, and so its total orbital quantum number $L = 3$. Its neighbor gadolinium, Gd, has seven 4f electrons and therefore $L = 0$. Because Gd has no spin–orbit coupling, it exhibits no magnetocrystalline anisotropy.

10.1.2 Symmetry of magnetocrystalline anisotropy

The symmetry of the magnetocrystalline anisotropy is always the same as that of the crystal structure. As a result, in iron, which is cubic, the anisotropy energy, E, can be written as a series expansion of the direction cosines α_i, of the saturation

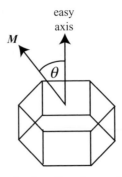

easy
axis

M

θ

Figure 10.4 Angle between magnetization direction and easy axis in a hexagonal material such as cobalt.

magnetization relative to the crystal axes:

$$E = K_1\left(\alpha_1^2\alpha_2^2 + \alpha_2^2\alpha_3^2 + \alpha_3^2\alpha_1^2\right) + K_2\left(\alpha_1^2\alpha_2^2\alpha_3^2\right) + \cdots. \tag{10.1}$$

Here K_1, K_2, etc. are called the *anisotropy constants*. Typical values for iron at room temperature are $K_1 = 4.2 \times 10^5$ erg/cm^3 and $K_2 = 1.5 \times 10^5$ erg/cm^3. The energy E is that stored in the crystal when work is done against the anisotropy 'force' to move the magnetization away from an easy direction. Note that the anisotropy energy is an even function of the direction cosines, and is invariant under interchange of the α_is among themselves.

Cobalt is hexagonal, with the easy axis along the hexagonal (c) axis. The anisotropy energy is uniaxial and its angular dependence is a function only of the angle θ between the magnetization vector and the hexagonal axis (see Fig. 10.4).

In this case the anisotropy energy can be expanded as

$$E = K_1 \sin^2\theta + K_2 \sin^4\theta + \cdots. \tag{10.2}$$

Typical values of the anisotropy constants for cobalt at room temperature are $K_1 = 4.1 \times 10^6$ erg/cm^3, and $K_2 = 1.0 \times 10^6$ erg/cm^3. Note that, in all materials, the anisotropy decreases with increasing temperature, and near T_C there is no preferred orientation for domain magnetization.

10.2 Shape anisotropy

Although most materials show some magnetocrystalline anisotropy, a polycrystalline sample with no preferred orientation of its grains will have no *overall* crystalline anisotropy. However only if the sample is exactly spherical will the same field magnetize it to the same extent in every direction. If the sample is not spherical, then it will be easier to magnetize it along a long axis. This phenomenon is known as *shape anisotropy*. Figure 10.5 shows the shape anisotropy constant as a function

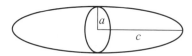

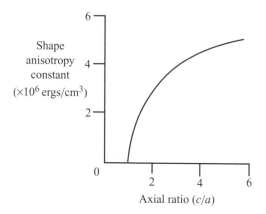

Figure 10.5 Shape anisotropy constant in a prolate spheroid of Co.

of the c/a ratio for a prolate spheroid of Co. Note that the anisotropy constant increases as the axial ratio increases, and that the shape anisotropy constant for typical axial ratios is of the same order of magnitude (around 10^6 ergs/cm^3) as the magnetocrystalline anisotropy constant.

In order to understand the origin of shape anisotropy, we first have to introduce the concept of the *demagnetizing field*.

10.2.1 Demagnetizing field

The concept of a demagnetizing field is confusing, and we will introduce it in a rather qualitative way from the viewpoint of magnetic poles. Let's suppose that our prolate spheroid from Fig. 10.5 has been magnetized by a magnetic field applied from right to left. This results in a north pole at the left end of the spheroid and a south pole at the right end. By definition, the lines of H radiate from the north pole and end at the south pole, resulting in the pattern of field lines shown in Fig. 10.6. We see from the figure, that the field *inside* the sample points from left to right – that is, in the opposite direction to the applied external field! This internal field tends to demagnetize the magnet, and so we call it the demagnetizing field, H_d.

The demagnetizing field is created by the magnetization of the sample, and in fact the size of the demagnetizing field is directly proportional to the size of the magnetization. We write

$$H_d = N_d M \tag{10.3}$$

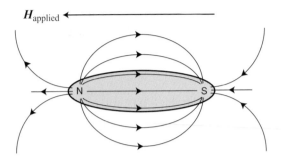

H_{applied}

Figure 10.6 **H** field around a prolate spheroid.

where N_d is called the demagnetizing factor, and depends only on the *shape* of the sample. Although we won't go into the details here, N_d can be calculated for different shapes (for details, see the book by Cullity[34]). The results of the calculations indicate that, for elongated samples, N_d is smallest along the long axis and largest along the short axis. The anisotropy becomes stronger as the aspect ratio increases, with $N_d \to 0$ as the distance between the 'poles' $\to \infty$.

Moreover, the effective field acting inside the material, H_{eff}, is smaller than the applied field by an amount equal to the demagnetizing field, i.e.

$$H_{\text{eff}} = H_{\text{applied}} - H_d. \tag{10.4}$$

So along the long axis, where N_d is small,

$$H_{\text{eff}} = H_{\text{applied}} - N_d M \simeq H_{\text{applied}} \tag{10.5}$$

and most of the applied field goes into magnetizing the sample. By contrast, along the short axis N_d is large, so

$$H_{\text{eff}} = H_{\text{applied}} - N_d M \ll H_{\text{applied}} \tag{10.6}$$

and so most of the applied field goes into overcoming the demagnetizing field. As a consequence it is easier to magnetize the sample along the long axis. This uniaxial magnetic response of needle-shaped particles leads to their widespread use as the media in magnetic recording systems. We will discuss this application in detail in Chapter 11.

Demagnetizing factors can be very important, and a high field is required to magnetize a sample with a large demagnetizing factor, even if the material has a large susceptibility. As an example, consider a sphere of permalloy, which is a Ni–Fe alloy with a coercive field, $H_c = 2$ A/m and saturation magnetization, $M_s = 1.16$ T. For a sphere, $N_d = \frac{1}{3}$; therefore the demagnetizing field, $H_d = N_d M \mu_0$ (in SI units), has the value 3.08×10^5 A/m. So to saturate the magnetization of the sphere we actually need to apply a field which is 10^5 times that of the coercive field!

Note that published magnetization curves have usually been corrected for demagnetizing effects, so that they represent the properties of the material independently of the shape of the sample.

10.3 Induced magnetic anisotropy

As its name suggests, induced magnetic anisotropy is not intrinsic to the material, but is produced by a treatment (such as annealing) which has *directional* characteristics. There is a huge potential for engineering the magnetic properties using such treatments because both the magnitude of the anisotropy and the easy axis can be altered by appropriate treatments.

Most materials in which magnetic anisotropy can be induced are polycrystalline alloys. By definition, if the grains in a polycrystalline material have a preferred orientation (which we call a 'texture') then there will be anisotropy. Preferred orientations are determined in part by the laws of physics (which we can't change), but also by the sample preparation. So some control over the degree and direction of preferred orientation is usually possible, using techniques such as casting, rolling or wire drawing. For the remainder of this chapter we'll discuss two methods in detail – magnetic annealing and roll anisotropy, and mention a few others.

10.3.1 Magnetic annealing

The term magnetic annealing refers to the heating and slow cooling of a sample in the presence of a magnetic field. In metal alloys this creates an easy axis of magnetization parallel to the applied field. The phenomenon was first observed in permalloy in the 1950s. Schematic hysteresis curves are shown in Fig. 10.7 for permalloy cooled in a field oriented parallel (Fig. 10.7(a)) and perpendicular (Fig. 10.7(b)) to the subsequently applied measurement field. It is clear that the

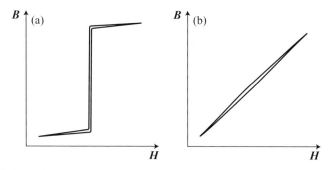

Figure 10.7 Schematic hysteresis loops of permalloy after annealing and cooling (a) in a longitudinal field and (b) in a transverse field.

observed hysteresis behavior can be attributed to uniaxial anisotropy induced with
the easy axis parallel to the annealing field.

Magnetic annealing induces anisotropy because it causes *directional order*. The
details of the physics are not understood, but we'll look at a schematic after we've
discussed roll anisotropy in the next section.

10.3.2 Roll anisotropy

A large magnetic anisotropy can also be created by *cold-rolling* Fe–Ni alloys. For
example, isoperm[TM], which is a 50:50 Fe–Ni alloy, can be cold-rolled with the (001)
plane in the sheet and [100] as the rolling direction (this is conventionally written
as (001)[100]). After recrystallization then subsequent rolling to 50% thickness
reduction, a large uniaxial anisotropy is created, with the easy axis in the plane of
the sheet and perpendicular to the rolling direction. As a result, subsequent magnetization *parallel* to the rolling direction takes place entirely by domain rotation,
giving a linear **B**–**H** curve, and a roughly constant permeability over a wide range
of applied fields. The geometry and magnetization curve are shown in Fig. 10.8.

10.3.3 Explanation for induced magnetic anisotropy

Both magnetic annealing and cold-rolling induce magnetic anisotropy because they
cause *directional order*. Iron and nickel atoms are able to migrate (particularly along
defects such as slip planes) so that, instead of forming a random solid solution,
there are an increased number of Fe–Fe or Ni–Ni neighbors along the direction

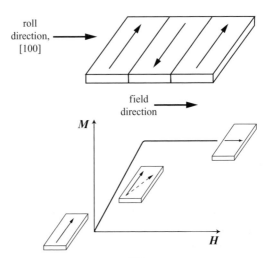

Figure 10.8 Cold-rolling of isoperm[TM] and resulting magnetization curve.

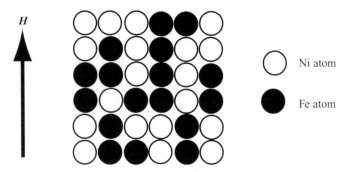

Figure 10.9 Directional order caused by magnetic annealing or cold-rolling.

of the applied field (in magnetic annealing) or perpendicular to the roll direction (in cold-rolling), as shown in Fig. 10.9. The details of why this ordering occurs, and why it results in a magnetic easy axis, are not understood, but it is believed to derive from the spin–orbit interaction.

10.3.4 Other ways of inducing magnetic anisotropy

If a Ni–Fe alloy is bombarded with neutrons in the presence of a magnetic field along the [100] direction, anisotropy is induced with the easy axis parallel to [100] and the hard axis parallel to [110]. Such *magnetic irradiation* creates defects which allow directional ordering to occur. Similarly, photo-induced magnetic anisotropy can be caused by annealing in the presence of electro-magnetic radiation, and stress annealing also causes anisotropy. As a result of the large variety of phenomena that can cause anisotropy, many thin-film magnetic alloys are anisotropic.

Homework

Exercises

10.1 Sketch the domain structure you would expect in *spherical* samples of ferromagnetic materials with the following properties:

- zero magnetocrystalline anisotropy
- large uniaxial anisotropy
- large magnetostriction
- a very small sample

10.2 What characteristics would you expect in the hysteresis loop of a ferromagnetic particle with average magnetocrystalline anisotropy, which is so small that it consists of a single domain? Suggest an application.

11

Magnetic data storage

"Magnetoresistance in metals is hardly likely to attract attention except in rather pure materials at low temperatures."

Sir A.B. Pippard, F.R.S. from *Magnetoresistance in metals.*
Cambridge University Press, 1989.

11.1 Introduction

The data storage industry is huge. Its revenue was tens of billions of United States dollars per year at the end of the 20th century, with hundreds of millions of disk, tape, optical and floppy drives shipped annually. It is currently growing at a rate of about 25 per cent, and the growth rate can only increase as the storing and sending of digital images becomes more common, with the phenomenal expansion of the world wide web and in ownership of personal computers.

Magnetic data storage is widely used in such applications as audio tapes, video cassette recorders, computer hard disks, floppy disks and credit cards, to name a few. Of all the magnetic storage technologies, magnetic hard disk recording is the most widely used. In this chapter, our main focus will be on the technology and materials used in writing, storing and retrieving data on magnetic hard disks. Along the way we will introduce some new phenomena, such as magnetoresistance and single-domain magnetism in small particles, and some new materials, both of which might have a large impact on storage technologies in the future.

The first hard disk drive (called RAMAC) was made by International Business Machines Corporation (IBM) in 1956. Its areal density (the number of bits per unit area of disk surface) was $2000\,\mathrm{bit/in^2}$, and the rate at which data were read or written was 70 kbit/s. Fifty 24-inch diameter disks were needed to hold five megabytes (MB) of data, and the size was similar to that of a large refrigerator. The cost was around $100,000 (or $20 per MB) and in fact storage space was usually leased rather than purchased. At press time (2002) a typical hard disk drive has an areal density

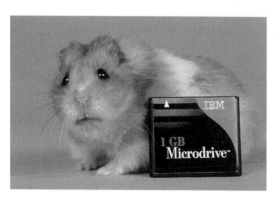

Figure 11.1 The 1-GB Microdrive™. Courtesy of International Business Machines Corporation. Unauthorized use not permitted.

of 15 Gbit/in² with more than three orders of magnitude improvement in the data transfer rate. Three or four 2.5-inch disks hold 60 gigabytes (GB) of data at a cost of around a hundred dollars (close to 1 cent per MB). And for the portable electronics market, IBM offers a 1-GB Microdrive™, which is smaller than a matchbook, weighs less than an ounce, and costs less than $500. Photographs of the 1 GB IBM Microdrive™ are shown in Fig. 11.1. On the left a hamster is posing next to the packaged Microdrive™, and on the right the open Microdrive™ is photographed next to a U.S. quarter-dollar for a size comparison. Again the areal density is 15 Gbit/in². In addition to conventional computer hard disk drives, applications for the Microdrive™ include digital cameras, handheld PCs, personal digital assistants, portable Internet music players and video cameras.

This huge decrease in cost per megabyte has been fueled in part by market forces – higher volume production and stiffer competition both lead to reduced costs – but also by improvements in materials. In particular, continually increasing areal densities allow more data to be stored for the same packaging and processing effort, and costs are reduced proportionally. The trend in areal density since 1985 is shown in Fig. 11.2, along with the development of one specific component – the read element in the recording head – that has facilitated the increase.

A photograph of the inside of a hard disk drive is shown in Fig. 11.3. A magnetic data storage system consists of three main components. First, the *storage medium* is the tape or disk in which the data is actually stored in the form of small magnetized areas. In the photograph in Fig.11.3, this is the large silver disk. The *write head* consists of a wire coil wound around a magnetic material which generates

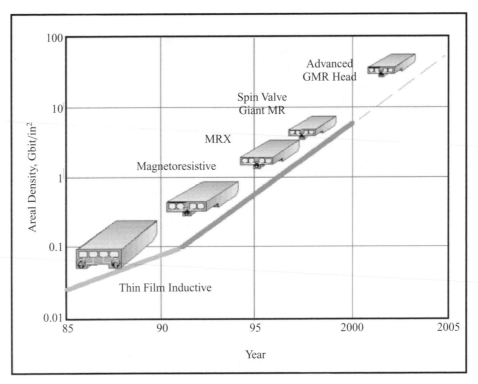

Figure 11.2 Past and projected increases in areal density (log scale) with time. The inserts show the materials used as the read head, see Section 11.4. Courtesy of International Business Machines Corporation. Unauthorized use not permitted.

a magnetic field (by electromagnetic induction) when current flows through the coil. This magnetic field writes the data by magnetizing the small data bits in the medium. Finally the *read head* senses the recorded magnetized areas, either by the reverse of the electromagnetic induction used in the write process, or using magnetoresistance – that is the change in resistance of a material when a magnetic field is applied. In the photograph, the read and write components are located in the recording head, at the end of the arm which moves over the disk. Clearly the material properties of the three components are linked, and there are many magnetic material design issues in the development of an entire magnetic storage device. For example higher areal densities are achieved by using higher-coercivity materials in the media (to stabilize smaller bits), lower head–disk spacings and more sensitive read heads (so that the field lines from the smaller bits can still be detected), and higher-magnetization write heads (to enable writing in the higher coercivity media).

For the remainder of this chapter we will discuss in turn the materials issues involved in the design and production of modern storage media, read heads and write heads. Along the way we'll also learn about the magnetic properties of small

Figure 11.3 The inside of a hard disk drive. Copyright 1998–2002 Seagate Technology LLC. Reproduced with permission of Seagate Technology LLC.

particles (which are used as the magnetic component in storage media) and about the phenomenon of magnetoresistance, which is widely used in the magnetic sensors in recording heads.

11.2 Magnetic media

The disk in a hard disk drive consists of four layers – a substrate, an underlayer, a magnetic layer where the data is actually stored, and a protective overcoat. Although the material properties of all the layers are relevant in determining the performance of the media, we will consider the magnetic layer first since it is the most important in our study of magnetic materials.

A primary requirement for the magnetic material used in the media is that it should have a large square hysteresis loop. Large loops occur in materials with large permeabilities and large coercivities. A large permeability is desirable because it results in strong lines of flux around each stored data bit, making the bits easier to detect. A large coercivity allows for permanent stable storage. The square shape means that there are two distinct stable magnetization states, and that the magnetization reversal takes place at a well-defined field strength.

In magnetic media, the square hysteresis loop is achieved by the use of small particles, which have characteristically large coercivities, and well-defined switching between magnetization directions. Therefore before we discuss specific types of media materials we will spend some time describing some of the relevant properties of magnetic small particles.

11.2.1 Magnetic properties of small particles

The magnetic properties of small particles are dominated by the fact that below a certain critical size a particle contains only one domain. Remember from Chapter 7 that the width of a domain wall depends on the balance between the *exchange* energy (which prefers a wide wall) and the *magnetocrystalline anisotropy* energy which prefers a narrow wall. The balance results in typical domain wall widths of around 1000 Å. So qualitatively, we might guess that if a particle is smaller than around 1000 Å a domain wall simply can't fit inside it, resulting in a single-domain particle!

We can make a better estimate of the size of single-domain particles by looking at the balance between the magnetostatic energy and the domain wall energy (Fig. 11.4). A single-domain particle (Fig. 11.4(a)) has high magnetostatic energy but no domain wall energy, whereas a multi-domain particle (Fig. 11.4(b)) has

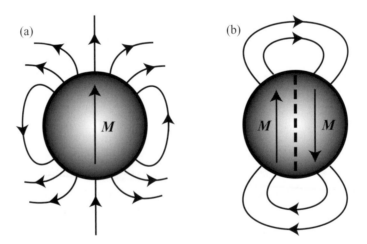

Figure 11.4 Balance between magnetostatic and domain wall energies in single- and multi-domain particles. (a) Single-domain particle with high magnetostatic energy. (b) Introduction of a domain wall reduces the magnetization energy but increases the exchange energy.

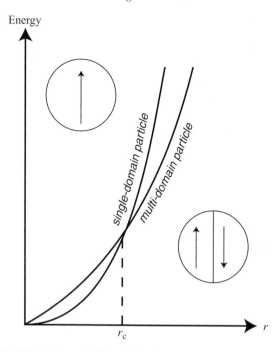

Energy

single-domain particle

multi-domain particle

r_c

r

Figure 11.5 Relative stability of single- and multi-domain particles.

lower magnetostatic energy but higher domain wall energy. It turns out that the reduction in magnetostatic energy is proportional to the volume of the particle (i.e. r^3, where r is the particle radius), and the increase in the domain wall energy is proportional to the area of the wall, r^2. The magnetostatic and exchange energies depend on particle radius as shown in Fig. 11.5. Below some critical radius, r_c, it is energetically unfavorable to form domain walls and a single-domain particle is formed.

Large single-domain particles can form if either the domain wall energy is large (because of, for example, large magnetocrystalline anisotropy) so that wall formation is unfavorable, or if the saturation magnetization is small, so that the magnetostatic energy is small.

Experimental evidence for single-domain particles

Small particles were known to have a large coercivity long before it was proved that they contain only one domain. The fact that the large coercivity of small particles is the result of single domains, rather than, for example, strain preventing easy domain wall motion was demonstrated in a seminal paper by Kittel and co-workers in the 1950s.[46] The authors made dilute suspensions of spherical Ni particles in paraffin wax, and measured the field required to saturate the magnetization of the

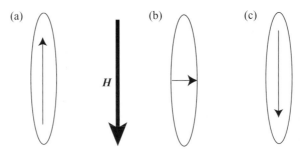

Figure 11.6 Magnetization mechanism in single-domain small particles.

samples for two particle diameters – 200 Å (below r_c) and 80,000 Å (above r_c). They found that the field required to saturate the small samples was 550 Oe, only slightly larger than that required to overcome the magnetocrystalline anisotropy. As a result they concluded that the particles consisted of single domains. By contrast, the field required to saturate the large particles was 2100 Oe, which is slightly higher than the demagnetizing field for Ni. This clearly indicates that the magnetization in the large particles takes place via a different mechanism (in this case domain wall motion and rotation). The large multi-domain particles can be kept in a saturated state only by a field larger than the demagnetizing field, whereas the small single-domain particles are always saturated, with the spontaneous magnetization in the same direction throughout their volume. The applied magnetic field required to magnetize a single-domain particle must overcome the anisotropy, but not a demagnetizing field.

Magnetization mechanism

Before application of an external field, the magnetization of a single-domain particle lies along an easy direction (as shown in Fig. 11.6(a)) which is determined by the shape and magnetocrystalline anisotropies. When an external field is applied in the opposite direction, the particle is unable to respond by domain wall motion, and instead the magnetization must rotate through the hard direction, (Fig. 11.6(b)) to the new easy direction (Fig. 11.6(c)). The anisotropy forces which hold the magnetization in an easy direction are strong, and so the coercivity is large. As we stated at the beginning of the chapter, this large coercivity is desirable for magnetic media applications.

Another desirable feature is the square hysteresis loop which results when the magnetic field is applied along an easy direction. Two stable states of opposite magnetization exist, and the field required to switch between them is well defined. A typical hysteresis loop for fields applied parallel to the easy direction is shown in Fig. 11.7(a). If the field is applied along a hard direction, there is initially no component of magnetization along the field direction. The field rotates the magnetization

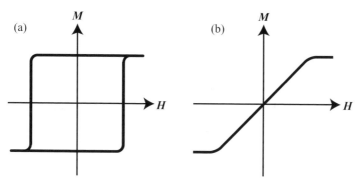

Figure 11.7 Hysteresis behavior of small particles with the external field applied parallel to (a) an easy direction, and (b) a hard direction.

into the field direction, but as soon as the field is removed it rotates back into the easy direction. As a result there is no hysteresis and the *B*–*H* curve is approximately linear, as shown in Fig. 11.7(b). Therefore for storage media, the particles must all be aligned with their easy axes parallel to the direction in which the write field will be applied. Any deviation from perfect alignment results in a loss of squareness of the overall *B*–*H* curve for the sample.

11.2.2 Materials used in magnetic media

There are two main types of media in production today both of which are composed of single-domain magnetic particles – particulate media and thin-film media. Typical coercive fields are around 3000 Oe, and typical grain sizes in the finer-grained thin-film media are tens of nanometers.

Particulate media

Particulate media consist of small, needle-like particles of, for example iron ferrite, γ-Fe_2O_3, or chromium oxide, CrO_2, bonded to a metal or polymer disk. The needles are aligned by a magnetic field during manufacture with their long axes parallel to the direction of motion past the read/write head. Each particle is a single domain that may be magnetized only with its moment aligned along the long axis (because of the shape anisotropy). The two binary data storage states correspond to (1) a change in magnetization between adjacent regions, and (0) no change in magnetization direction.

Iron oxide particles have been traditionally very widely used because they are chemically stable, pollution free and inexpensive. γ-Fe_2O_3 and Fe_3O_4 can be prepared by dehydration, oxidation or reduction of acicular α-FeOOH, which results in needle-like particles 0.3–0.7 μm long, and $\sim$0.05 μm in diameter. A higher H_c

can be obtained using Co-modified iron oxide particles, consisting of a core of Fe_3O_4 coated epitaxially by cobalt ferrite.

Thin-film media

A problem with particulate media is that voids can disrupt the homogeneous distribution of particles, resulting in less uniform orientation and a lower coercivity. The thin-film arrangement allows for higher storage density than is possible in particulate media because the packing efficiency is much higher. Thin film media consist of approximately 10–50 nm thick polycrystalline magnetic alloys such as CoPtCr or CoCrTa deposited on a substrate. This time each 10–30 nm-sized grain within the film is a single domain. The primary magnetic component is the Co, and the purpose of the Pt or Ta is to increase the coercivity by increasing the anisotropy. The Cr segregates to the grain boundaries and so reduces the drop in coercivity caused by undesirable intergranular exchange (see discussion in the next section). The crystallographic direction of easy magnetization (in this case produced by the magnetocrystalline anisotropy) is aligned along the direction of the disk motion.

11.2.3 Disadvantages of small particles in media applications

In spite of their desirable hysteresis properties, there are two problems associated with using small particles for the magnetic media in storage devices. The first is the detrimental effect of inter-particle interactions, and the second is a reduction in coercivity which occurs at very small particle sizes (so-called 'superparamagnetism'). We discuss each of these effects here, and the efforts which are being made to overcome them for technological applications.

Inter-particle interactions

It has been observed experimentally that when the anisotropy of small particles is derived primarily from shape anisotropy, the coercive field drops as the packing density is increased. This is a result of inter-particle interactions. Qualitatively we can understand this effect by considering the field that a magnetized particle exerts on its neighbor, as shown in Fig. 11.8. All the particles are initially magnetized in the up direction. We see from the figure that the field which particle A exerts on particle C acts in the down direction. So when the external field is reversed and applied in the down direction, the field from particle A acting on particle C assists the applied external field, and so C reverses its magnetization at a lower applied field than it would in isolation. Overall the sample has a lower coercivity than a collection of isolated particles. (Of course we can see from the picture that the opposite effect occurs at B – that is, the field from A works *against* the external reversed field.

Coercivity
(Oe)

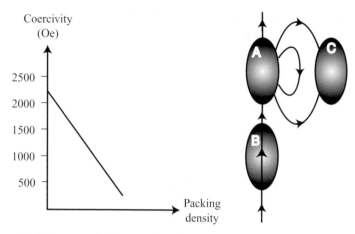

Figure 11.8 Inter-particle interactions in media composed of small particles.

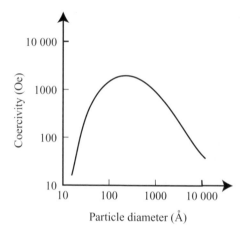

Figure 11.9 Coercivity as a function of size for small particles.

In reality the situation is more complicated than our simple picture!) As the packing density increases the interactions increase, therefore for media applications, where a high packing density is desirable for reduced signal-to-noise ratio, it is necessary to isolate the particles, by for example precipitation of non-magnetic material along the boundaries of the magnetic grains.

Superparamagnetism

Figure 11.9 is a schematic of the variation in coercivity with particle diameter. As the size of the sample is reduced from the bulk, the coercivity initially increases as single-domain particles are formed. Below some critical radius, however, the coercivity decreases and eventually drops to zero.

The drop in coercivity at very small particle size is the result of a corresponding reduction in anisotropy energy with size. The anisotropy energy, which holds the magnetization along an easy direction, is given by the product of the anisotropy constant, K, and the volume of the particle. As the volume is reduced, KV becomes comparable to the thermal energy $k_B T$. As a result thermal energy can overcome the anisotropy 'force' and spontaneously reverse the magnetization of a particle from one easy direction to the other, even in the absence of an applied field.

This phenomenon is called 'superparamagnetism' because, as a result of this competition between anisotropy and thermal energies, assemblies of small particles show behavior similar to that of paramagnetic materials, but with a much larger magnetic moment. The magnetic moment of a 50 Å particle is typically around 10,000 μ_B, whereas that of a magnetic atom is of the order of the Bohr magneton. In both cases an applied field tends to align the magnetic moments, and thermal energy tends to disalign them. However, because the magnetic moment in super-paramagnetic particles is so much larger than that in atoms, the particles become aligned at correspondingly smaller values of the magnetic field.

If the anisotropy is zero, then the magnetic moment of each particle can point in any direction, and the classical theory of paramagnetism reproduces the behavior of the particles well. The magnetization is described by the Langevin function, as we saw in Section 5.1:

$$M = Nm \left[\coth \left(\frac{mH}{k_B T} \right) - \frac{k_B T}{mH} \right] \tag{11.1}$$

$$= NmL(\alpha), \tag{11.2}$$

where $\alpha = mH/k_B T$ and $L(\alpha) = \coth(\alpha) - 1/\alpha$ is called the Langevin function. In this case, however, because the magnetic moment per particle, m, is large, α is correspondingly large, and so the full magnetization curve, up to saturation, can be observed easily even at moderate fields. (Remember that for ordinary paramagnetic materials, very high fields and low temperatures were required to study the whole magnetization curve).

If the anisotropy of each particle is finite, and the particles are aligned with their easy axes parallel to each other and the field, then the moment directions are quantized, with two allowed orientations. In this case the magnetization is described by the special case of the Brillouin function with $J = \frac{1}{2}$, i.e.

$$M = Nm \tanh(\alpha). \tag{11.3}$$

Again the entire magnetization curve can be obtained even at moderate fields.

In the general case the particles are not perfectly aligned, and neither of these ideal equations exactly describes the observed magnetization curve. Also, in most samples the particles are not all the same size and the moment per particle is

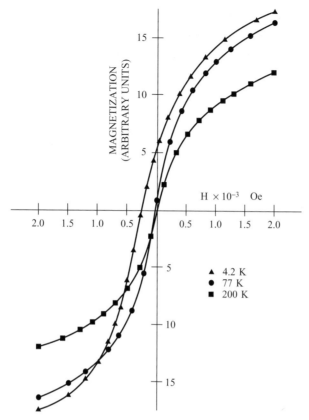

Figure 11.10 Magnetization curves of iron particles above and below the superparamagnetic transition temperature from Ref. 47. Copyright (1956) American Institute of Physics. Reproduced with permission.

not constant, giving further deviation from ideality. In all cases however, there is no hysteresis (that is the coercivity and the remanent magnetization are both zero), and so superparamagnetic materials are not suitable for recording media. Superparamagnetism can be destroyed by reducing the temperature, increasing the particle size, or increasing the anisotropy, such that $KV > k_B T$.

Figure 11.10 shows the magnetization curves of 44 Å diameter iron particles from one of the earliest studies of superparamagnetism.[47] At 200 K and 77 K the particles show typical superparamagnetic behavior with no hysteresis. Note also that the induced magnetization is higher at 77 K than at 200 K for the same applied field, as we would expect from the Langevin theory. At 4.2 K, however, the particles do not have enough thermal energy to come to equilibrium with the field and so hysteresis is observed (only half of the hysteresis loop is shown in the figure). An operating temperature of 4.2 K is clearly undesirable in a practical device, and so these particles would not be suitable for magnetic media applications!

11.2.4 The other components of magnetic hard disks

In addition to the magnetic layer, magnetic hard disks also contain a substrate, an underlayer and an overcoat. The requirements for the substrate layer are high hardness and low density for shock resistance (this is particularly important in lap-top computers), high modulus for reduced vibration, good thermal stability for stability during processing, absence of defects, and low cost. Traditionally an Al–Mg alloy plated with $\sim$10 μm of NiP was used, but more recently there has been a transition to glass substrates. The choice of substrate greatly affects the subsequent processing and performance of the disk. For example the nucleation and growth of the underlayer are different on glass than on NiP, which in turn affects the grain size and crystallographic orientation of the magnetic layer.

The purpose of the underlayer is to control the crystallographic orientation and grain size of the magnetic layer, to promote adhesion, to protect the substrate from corrosion, and to physically isolate the magnetic grains from each other, in order to prevent the problems with inter-particle interactions referred to above. The current material of choice for the underlayer is chromium, which tends to orient the easy axis of Co (used in the magnetic layer) in the plane of the film. Cr alloys such as CrV are also used in order to improve the lattice matching between the underlayer and the magnetic layer.

Finally, the overcoat serves to prevent wear of the magnetic layer and subsequent data loss during contact with the head. It also provides a low friction interface between the magnetic layer and the head. The material of choice is an amorphous three-dimensional C:H film covered by a lubricant of perfluoropolyether.

11.3 Write heads

In magnetic hard disks, writing is achieved by the process of electromagnetic induction. A magnetic field produced by a current circulating in the write head intersects with the media and magnetizes it, creating a data bit. A schematic of a write head is shown in Fig. 11.11.

The purpose of the magnetic material around which the wire is wound is to concentrate the magnetic flux generated by the current passing through the wire. The gap between the write poles allows some of the flux to leak out, creating 'fringing fields' which actually magnetize the media. The magnetic material in the write head should have a large permeability, so that large magnetic fields are generated, and a low coercivity, so that its direction of magnetization can be easily reversed. Traditionally write heads were made of cubic ferrites, which are soft, and therefore easily magnetized. However the saturation magnetizations are not large, so strong magnetic fields can not be generated. In modern heads, metals

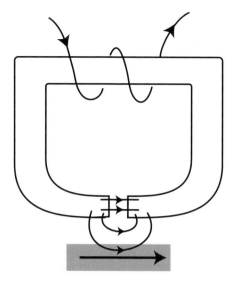

Figure 11.11 Schematic of an inductive write head.

such as permalloy, which has a higher saturation flux density, are used. The higher saturation flux density facilitates writing in higher-coercivity media, and allows for narrower track widths and in turn a higher storage density. However, modern data rates are so high that eddy currents are induced in metal heads, which limit the operation frequency. Therefore there is a move to laminated thin-film heads of, for example, FeAlN, in which the eddy currents are suppressed allowing for an improved high-frequency response. FeAlN thin films are soft, with a coercivity of less than 1 Oe, they have a saturation magnetization of 20 kG and a permeability of 3400 with approximately zero magnetostriction. A 4-Gbit/in^2 inductive write head was produced in the mid-1990s using poles made of 2 μm thick multilayer FeAlN/SiO$_2$, subsequently trimmed to 1 μm track width using focused ion beam etching.[48] The heads were able to magnetize 2950 Oe coercivity media without undergoing saturation. Incremental improvements in design have allowed today's recording densities to reach ~10 Gbit/in^2. New materials with even higher permeabilities and higher resistivities, such as CoZrCr, are being explored for future higher density and faster data rate applications.

11.4 Read heads

In the past, the same inductive component that performs the write operation was also used for the read head. This had the obvious advantage of reducing the number of components contained within the recording head. However the fields emanating from a stored data bit are small, therefore the signals generated in the read head using electromagnetic induction were correspondingly weak. Today a separate component, which is based on the phenomenon of magnetoresistance rather than magnetic induction, is used to detect the stored data bit.

11.4.1 Magnetoresistance – general

The term magnetoresistance refers to the change in resistance of material when a magnetic field is applied. The magnetoresistive (MR) ratio is defined to be the ratio of the change in resistance when the field is applied to the resistance at zero field, that is

$$\text{MR ratio} = \frac{R_H - R_0}{R_0}$$
$$= \frac{\Delta R}{R}.$$

So a material which has a larger resistance in the presence of a field than in the absence of a field is defined to have a positive magnetoresistance, whereas if the field reduces the resistivity, the magnetoresistance is negative.

The phenomenon of magnetoresistance has been the subject of a great deal of recent research interest, and magnetoresistive materials are used today in a number of commercially available technologies, such as magnetic sensors, magnetic recording heads, and magnetic memories. In this section we will describe the nature and origin of magnetoresistance in normal metals, then discuss anisotropic magnetoresistance in ferromagnetic metals, giant magnetoresistance in metallic multilayers, and colossal magnetoresistance in perovskite structure manganites. Along the way we'll look at the structure of some of the read heads made from various magnetoresistive materials.

11.4.2 Magnetoresistance in normal metals

In the absence of an external field, electrons travel through a solid in straight lines in between scattering events as shown below:

For a free-electron gas, the same is true even in the presence of an applied field. Although the applied field exerts a force (the Lorentz force) on the electrons which deflects them from their path, the electric field created by the displaced electrons exactly balances the Lorentz force, and at equilibrium the electrons follow the same straight-line path. This is known as the Hall effect, and is illustrated in Fig. 11.12.

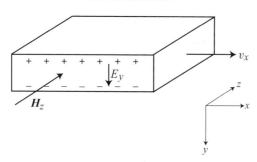

Figure 11.12 Hall effect in a free electron gas.

Here the electrons moving with velocity v in the x direction are initially deflected towards the y direction by a field H applied in the z direction. As a result of the exact balance between this Lorentz force and the induced electric field, E_y, an ideal free electron gas will have zero magnetoresistance.

However in a 'real' metal, the conduction electrons have different mean velocities, and although on average the transverse Hall electric field exactly balances the magnetic field, individual electrons travel in a curved path as shown below:

Since the Lorentz force, $ev \times B$, curls the electrons into orbits, they travel further and scatter more, and so the resistance in the presence of the field is larger than the resistance in the absence of the field. Therefore the magnetoresistance in normal metals is positive. The effect is, however, very small, and does not have a technological application.

11.4.3 Magnetoresistance in ferromagnetic metals

Anisotropic magnetoresistance

Larger magnetoresistive effects, of around 2%, are observed in ferromagnetic metals and their alloys. The phenomenon is called *anisotropic magnetoresistance* (AMR) because the change in resistance when a field is applied parallel to the current direction is different from that when the field is perpendicular to the current direction. In fact the resistance for current flowing parallel to the field direction, $\rho_{parallel}$, increases when a field is applied, whereas the resistance for current flowing perpendicular to the field direction, $\rho_{perpendicular}$, decreases by approximately the same amount. The effect is significant even in small fields, and the magnetoresistance saturates at 5–10 Oe. A schematic of the resistance changes in an AMR material is shown in Fig. 11.13. The origin of AMR lies in the spin–orbit coupling, and was first explained by Kondo in the early-1960s.[49] The s electrons which are responsible for the conduction are scattered by the unquenched part of the orbital

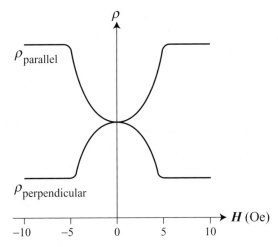

Figure 11.13 Anisotropic magnetoresistance in a ferromagnetic metal such as permalloy for field applied parallel and transverse to the current direction.

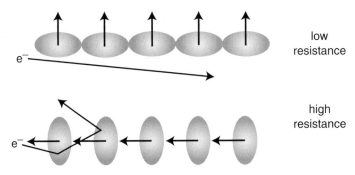

Figure 11.14 The origin of AMR.

angular momentum of the 3d electrons. There is experimental evidence supporting this assumption in the fact that the observed magnetoresistance correlates with the deviation of the gyromagnetic ratio from its spin-only value of 2. As the magnetization direction rotates in response to the applied magnetic field, the 3d electron cloud deforms, and changes the amount of scattering of the conduction electrons. The process is shown schematically in Fig. 11.14; when the magnetization direction is perpendicular to the current direction, the scattering cross-section is reduced compared with the zero-field case, whereas when the magnetization direction is parallel to the current direction the scattering cross-section is increased.

AMR read heads

From about 1993 until the late-1990s, anisotropic magnetoresistive materials were used almost exclusively as the read elements in recording heads. A typical head design, the so-called dual stripe recording head, is shown in Fig. 11.15. The current

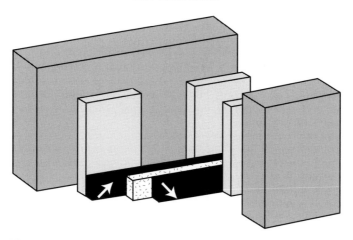

Figure 11.15 Schematic of a dual stripe magnetoresistance head.

flows from the current leads (light gray) along the lengths of the AMR bars (black), which are separated by a thin dielectric layer (speckled). The dark-gray bars are shields to reduce the effects of stray fields. The dual stripe design utilizes the transverse magnetoresistance, with the current running perpendicular to the field. The magnetic fields generated by the current in one stripe bias the other stripe, and vice versa, resulting in a linear signal.

Magnetoresistance from spontaneous magnetization

In normal non-magnetic metals, the resistivity decreases smoothly with decreasing temperature. This is the result of decreased thermal vibrations of the atoms causing a more ordered lattice, in turn causing less scattering of the conduction electrons. Below the ferromagnetic ordering temperature in ferromagnetic metals there is an additional reduction in the resistivity, beyond that which is observed in normal metals. This additional reduction in resistivity is due to the increased *directional* ordering of the magnetic moments, which also results in less scattering of the conduction electrons.[37]

Giant magnetoresistance

The magnetoresistive component in modern read heads operates on the so-called *giant* magnetoresistive (GMR) effect. This effect was first observed in 1988[50] in antiferromagnetically coupled metallic multilayers of Fe/Cr. Data from the original paper are shown in Fig. 11.16. Typical magnetoresistance values are an order of magnitude larger than those seen in AMR materials.

In GMR materials, thin layers of magnetic material are separated by layers of non-magnetic material. Depending on the thickness of the non-magnetic layers, the magnetic layers couple either ferromagnetically or antiferromagnetically.

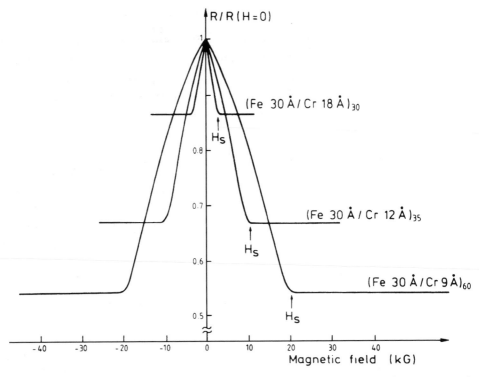

Figure 11.16 Magnetoresistance of three Fe/Cr superlattices at 4.2 K. The current and the applied field are along the same [110] axis in the plane of the layers. From Ref. 50. Copyright (1988) by the American Physical Society.

(Although the details of the coupling mechanism are not known, an RKKY-like mechanism predicts results which are similar to experimental observations). Giant magnetoresistance occurs when the thicknesses are chosen such that the adjacent layers are antiferromagnetic in zero applied field, as shown in Fig. 11.17(a). This results in a high-resistance state as up-spin electrons are scattered by regions of down-spin magnetization and vice versa. Then the GMR effect works by changing the relative magnetization directions between adjacent magnetic layers. A low-resistance state is obtained when a magnetic field strong enough to overcome the antiferromagnetic coupling is applied and the magnetization of the layers is rotated to a ferromagnetic configuration, as shown in Fig. 11.17(b). When the magnetic layers are ferromagnetically aligned, conduction electrons of compatible spin-type are able to move through the heterostructure with minimal scattering, and the overall resistance of the material is lowered.

The difference in scattering between antiferromagnetically and ferromagnetically aligned multilayers can be understood within a band structure picture.[51] As shown schematically in Fig. 11.18, in a normal metal there are equal numbers of up- and

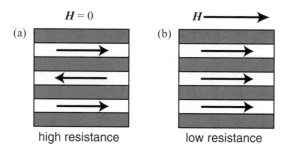

Figure 11.17 Schematic of the high- and low-resistance states of GMR multilayer systems.

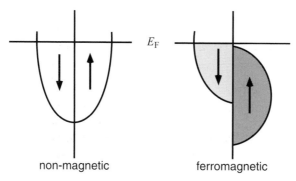

Figure 11.18 Schematic densities of states in a normal metal (*left*) and in a half-metallic ferromagnet (*right*).

down-spin states at the Fermi level, therefore up- and down-spin electrons travel through a normal metal with equal probability. In a spin-polarized metal however, there are more states of one spin direction than the other at the Fermi level. In the particular example shown in Fig. 11.18, only down-spin states are available at the Fermi level, and hence only down-spin electrons can travel through the system. Such a material is said to be *half-metallic*, since it is metallic for one spin polarization and insulating for the other. Provided that adjacent magnetic layers are magnetized in the same direction, the down-spin electrons are able to conduct through the system, since down-spin states continue to exist at the Fermi level. Therefore the ferromagnetic arrangement has a low resistance. If an adjacent layer is aligned *antiferromagnetically*, however, the up- and down-spin densities of states are reversed, giving only up-spin states at the Fermi level. The down-spin electrons entering the layer have no states to occupy, and hence are scattered. Therefore the antiferromagnetic arrangement has a high resistance.

GMR heads – spin valves

The original data for GMR multilayers suggested that large fields, of the order of tens of kilogauss, were required to overcome the antiferromagnetic coupling

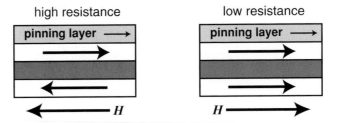

Figure 11.19 Schematic of the operation of a spin-valve system.

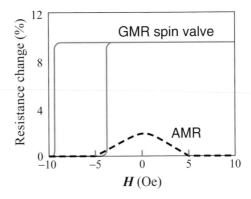

Figure 11.20 Comparison of the magnetoresistance in AMR and GMR spin-valve heads.

and rotate the magnetization to the ferromagnetic orientation. However a so-called *spin-valve* architecture has been developed in which films can be switched from antiferromagnetic to ferromagnetic at much lower fields. Typical magnetoresistances for spin valves are of the order of tens of per cent in fields of tens of oersteds. In addition spin valves have a uniform field response, which makes them much more appealing for use as sensors, for example in recording heads.

In spin valves two magnetic layers are separated by a non-magnetic spacer layer, as shown in Fig. 11.19. One of the magnetic layers has its magnetization direction pinned using exchange-bias coupling to an adjacent antiferromagnetic layer. The lower magnetic layer is free to switch back and forth in the presence of an applied magnetic field. Just as in GMR multilayers, spin-dependent scattering gives a low-resistance state when the magnetic layers are ferromagnetically aligned, and a high-resistance state in the antiferromagnetic configuration.

A comparison of the magnetoresistance obtained from AMR and GMR spin-valve heads is shown in Fig. 11.20. Note that the magnitude of the magnetoresistance is considerably larger in the GMR spin valve. Also we see that the hysteresis loop is shifted, as we explained in our discussion of exchange-bias coupling in Chapter 8.

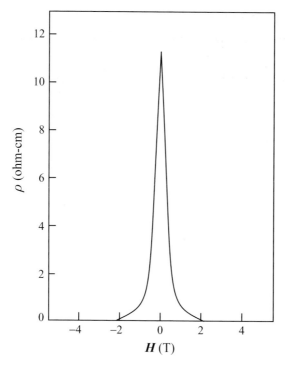

Figure 11.21 Typical change in resistance with applied field in a LaCaMnO film at 77 K.

11.4.4 Colossal magnetoresistance

Colossal magnetoresistance (CMR) was first observed in 1994 by Jin *et al.*[52] in the perovskite structure manganite, $La_{0.67}Ca_{0.33}MnO_3$. The term colossal was chosen because of the very large change in resistance, essentially from an insulating to a conducting state, on application of a magnetic field. A typical response in resistivity as a function of applied field is shown in Fig. 11.21. Although the original experiments were at low temperature, similar effects have since been observed at or near room temperature. However large fields, of the order of a few teslas, are still required to cause the change in resistance. Therefore CMR materials are not currently considered likely to find direct practical application as magnetic sensors, or in particular as the read element in recording heads. A number of other applications are being explored however, including their use in bolometers, where a change in temperature causes a change in conductivity driven by a metal–insulator transition, and in spin-tunneling devices that exploit their half-metallicity. Finally, it is possible that switching might be achieved at practical field strengths by using clever device architectures such as magnetic tunnel junctions.[53]

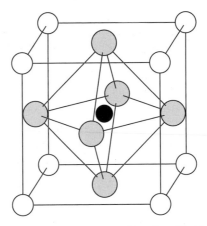

Figure 11.22 The perovskite structure. The small cation (in black) is surrounded by an octahedron of oxygen anions (in gray). The large cations (white) occupy the unit cell corners.

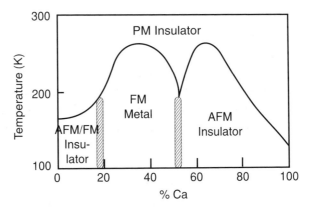

Figure 11.23 The phase diagram of $La_{1-x}Ca_xMnO_3$. From Ref. 56. Copyright (1995) by the American Physical Society.

Superexchange and double exchange

In order to interpret the properties of CMR materials, we first need to understand their structure in some detail. The perovskite structure (Fig. 11.22) consists of a small cation, manganese in this case, surrounded by an octahedron of oxygen anions, with a large cation, La or Ca here, filling the space at the corners of the unit cell. Note the O–Mn–O–Mn chains running along all three Cartesian directions. Perovskite structure manganites were studied extensively in the 1950s, in part because they have a very rich phase diagram, with both the magnetic and structural ordering depending on the amount of doping and the temperature.[54, 55] A modern phase diagram of the (La,Ca)MnO₃ system is shown in Fig. 11.23.[56]

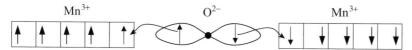

Figure 11.24 Superexchange between two empty Mn 3d orbitals, leading to antiferromagnetic coupling of the Mn magnetic moments.

We saw in Chapter 8 how the superexchange mechanism leads to antiferromagnetic coupling between filled transition metal d orbitals coupled by filled anion p orbitals. Likewise, *empty* transition metal d orbitals coupled by filled oxygen p states are antiferromagnetic. In this case, the oxygen acts as a ligand, donating electron density into the empty transition metal d state, and by Hund's rules, the donated electron should have the same spin as the electrons occupying the filled transition metal d states. This mechanism is shown in Fig. 11.24. Finally, by the same arguments, the anion-mediated coupling between one empty and one filled transition metal d orbital is ferromagnetic, since the empty orbital accepts an electron of the same spin type as the electrons contributing to its magnetic moment, and the filled orbital accepts one of opposite spin.

In one end-point CMR compound, $LaMnO_3$, the Mn^{3+} ions each have four d electrons. As a result, some oxygen anions join pairs of filled orbitals, and some join filled Mn d orbitals to empty Mn d orbitals. Geometric arguments[54] account for the experimentally observed A-type antiferromagnetism, in which (100) planes of ferromagnetically aligned Mn ions are coupled antiferromagnetically to each other. In the other end-point compound, $CaMnO_3$, the Mn^{4+} ions have only three d electrons each. It is then always possible to have ligand stabilization into empty Mn d orbitals, leading to antiferromagnetic interactions in all directions. The resulting structure is called G-type antiferromagnetic.

In mixed-valence compounds such as $La_xCa_{1-x}MnO_3$, an additional mechanism, proposed by Zener[57] and known as double exchange, influences the magnetic ordering. If an oxygen anion couples two Mn ions of different valence, such as a Mn^{3+} and a Mn^{4+} ion, then there are two possible configurations:

$$\psi_1: Mn^{3+} \; O^{2-} \; Mn^{4+}$$
$$\psi_2: Mn^{4+} \; O^{2-} \; Mn^{3+}$$

which have the same energy. If there is a finite probability that an electron initially on the Mn^{3+} ion is able to transfer to the Mn^{4+} ion (converting ψ_1 into ψ_2) then the degeneracy will be lifted, lowering the energy of one of the new states, and hence the overall energy, by the magnitude of the transfer matrix element. Such an electron transfer can only occur if the magnetic moments on the two Mn ions are parallel, and so the lower-energy state can only be obtained for ferromagnetic ordering.

The CMR effect is strongest in the region of Ca doping around $\frac{1}{3}$. We see that in this region, the material undergoes a phase transition from a paramagnetic insulating phase to a low-temperature ferromagnetic metal. Although the details of the CMR mechanism are still not entirely well understood, it is believed that an applied field causes a similar phase transition, with a corresponding increase in conductivity.

11.5 Future of magnetic data storage

The decrease in cost per bit discussed in Section 11.1 has been possible in large part because of a simultaneous increase in areal density. Therefore maintaining or improving existing areal density trends into the future is a high priority for disk drive manufacturers. In this section we outline some of the anticipated problems associated with a continued increase in areal density, and some solutions that are being explored to overcome them. A more detailed review of the future of magnetic data storage technology can be found in Ref. 58.

There are three primary impediments to the continuation of the trends shown in Fig. 11.2. The first two, the superparamagnetic limit, which we discussed in Section 11.2.3, and fundamental limitations in switching speed, result from the fact that the electrical and magnetic properties change as we scale down in size. A third obstacle, the reduction in head-to-disk spacing down to atomic dimensions, results from an incompatibility between further miniaturization and existing device architectures. Therefore the evolution of magnetic storage devices will undoubtedly, at least in the near term, follow two pathways – continued optimization of current disk design, which is in fact remarkably similar to the original hard disks of the 1950s, and the development of new recording architectures.

Currently each data bit needs to contain a few hundred magnetic particles, otherwise the signal-to-noise ratio for bit detection becomes unacceptable. Therefore, as the bit size is decreased, the particle size must decrease correspondingly. As we discussed earlier, below a certain critical size a magnetized particle can spontaneously switch magnetization direction, as the thermal energy exceeds the anisotropy energy pinning the magnetization in place. If no other aspects of the device were changed, loss of stability would occur at an areal density $\sim$40 Gbit/in^2. One option for achieving higher areal densities rests in the fact that the signal-to-noise ratio actually scales as the perimeter length of the bit, and so is determined by the aspect ratio as well as the surface area. Making narrower tracks, and consequently bits with a lower aspect ratio, would reduce the signal to noise and therefore allow larger particle sizes for the same bit size. Narrower tracks, however, are more prone to interference with their neighbors. A second option is the development of better error-correcting codes, which will allow lower signal-to-noise ratios to be tolerated. With these advances, densities of 100–200 Gbit/in^2 might be expected before the superparamagnetic limit is reached.

A change from conventional longitudinal recording, in which the magnetization direction is in the plane of the film, to perpendicular recording, where the magnetization is normal to the film surface, would undoubtedly allow even higher areal densities. The optimal medium thickness for perpendicular recording is thicker than that for longitudinal recording and so the grain volume can be larger. Also, since the geometry is more efficient, the write field at the medium is stronger. Perpendicular recording might further increase areal densities two to four times before the superparamagnetic limit is reached. However the significant change needed in architecture is expensive to make, and most existing studies are being performed by research institutions rather than magnetic storage device manufacturers.

A quite different avenue for exploration is the formation of media with only one magnetic grain per bit cell. This could be achieved for example by photolithographic methods, which have the disadvantage of being expensive, or by chemical synthesis of monodisperse magnetic nanoparticles, which is cheaper but challenging.[59] An additional increase in areal density of roughly an order of magnitude could be expected.

In terms of marketability, the next factor after cost and capacity is data rate. The data rate is in turn determined by the speed at which the head is able to switch the bits in the media. Magnetic switching times ~10 ns are now state of the art, and below this the magnetic properties of both heads and media start to change significantly. For example at higher switching rates even modern laminated heads are prone to the formation of eddy currents. A more fundamental limitation is that bits in the media take a few nanoseconds to flip once the field is applied, since the flipping process depends on damping of the precession induced by the applied field. The problem is compounded as the particle size approaches the superparamagnetic limit and the bits become less stable.

In conclusion, in spite of some fundamental physical difficulties, increasing trends in areal density and consequent cost reductions in magnetic data storage are likely to continue for the immediate future.

Homework

Exercises

11.1 Review question
   (a) Calculate the magnetic field generated by an electron moving in a circular orbit of radius 1 Å with angular momentum $\hbar$ J s, at a distance of 3 Å from the center of the orbit, and along its axis.
   (b) Calculate the magnetic dipole moment of the electron in (a). Give your answer (i) in SI and (ii) in cgs units.
   (c) Sketch the field lines around the magnetic dipole, when it is oriented such that its north pole is pointing upward. What would be the preferred orientation of a second

dipole be if it were (i) vertically above the original dipole (i.e. along its axis), or (ii) horizontal from the original dipole?

(d) Based on your answer to (c), sketch the magnetic ordering in a 3D lattice of magnetic moments, assuming that the classical dipole–dipole interaction is the principal driving force between the moments.

(e) Calculate the magnetic dipolar energy of an electron in the field generated by a second electron at a distance of 3 Å away along its axis, assuming that the magnetic moment of the second electron is aligned (i) parallel and (ii) antiparallel to the field from the first electron. Based on your answer, estimate the ordering temperature of your 3D lattice of classical magnetic moments.

(f) What are the electronic structures of Mn^{3+} and Mn^{4+} ions? What are the magnetic moments of these ions (assuming that only the spin and not the orbital angular momentum contributes to the magnetic moment)?

(g) Use chemical bonding arguments to predict the magnetic structure of a cubic 3D lattice of (i) Mn^{3+} ions linked by oxygen anions (such as is found in $LaMnO_3$) and (ii) Mn^{4+} ions linked by oxygen anions (such as is found in $CaMnO_3$). Given that the Néel temperature of $CaMnO_3$ is around 120 K, compare the strength of the Mn–Mn interactions in $CaMnO_3$ with those between the classical magnetic moments described above.

(h) What kind of magnetic interaction would you expect between two adjacent manganese ions, one of which is Mn^{3+} and one of which is Mn^{4+}, which are bonded by an O^{2-} ion? (Such an arrangement occurs in the colossal magnetoresistive material $La_{1-x}Ca_x MnO_3$.)

12

Magneto-optics and magneto-optic recording

"We are in great haste to construct a magnetic telegraph from Maine to Texas; but Maine and Texas, it may be, have nothing important to communicate."

Henry David Thoreau, *The writings of Henry David Thoreau, vol. 2,*
p. 58. Houghton Mifflin, 1906.

We begin this chapter with a discussion of the physics behind a phenomenon known as the *magneto-optic* (MO) effect, which, as its name implies, concerns the interaction of light with magnetic materials. Then we describe both the mechanism and the materials used in one specific application of magneto-optics – that of magneto-optic data storage.

12.1 Magneto-optics basics

The term 'magneto-optics' refers to the various phenomena which occur when electro-magnetic radiation interacts with magnetically polarized materials. Here we describe two important, and related magneto-optic phenomena, the Kerr effect and the Faraday effect.

12.1.1 Kerr effect

The Kerr effect is the rotation of the plane of polarization of a light beam during *reflection* from a magnetized sample. For most materials the amount of rotation is small (of the order of tenths of a degree) and depends on both the direction and magnitude of the magnetization. The Kerr effect can be used in observation of magnetic domains, as shown schematically in Fig. 12.1.

Radiation from a light source is first passed through a polarizer. The resulting plane-polarized light is then incident on a sample which, in our example, contains

159

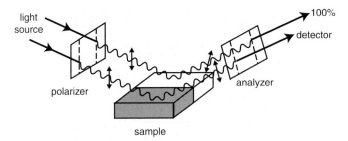

Figure 12.1 Domain observation using the Kerr effect. The gray and white regions of the sample correspond to domains of opposite magnetization.

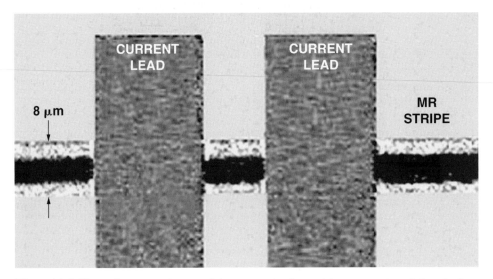

Figure 12.2 Kerr microscope image of the magnetic element in a magnetoresistive device. Reproduced with permission from Ref. 60. Copyright (1995) IEEE.

two domains magnetized in opposite directions. The light incident on one domain is rotated in the opposite direction from that incident on the other domain. Therefore if the analyzer is oriented such that the light reflected from the first domain is 100% transmitted, then the plane of polarization of the light reflected from the other domain is not aligned with the analyzer, and the transmittance is reduced.

Two different examples of images recorded using the magneto-optic Kerr effect are shown in Figs. 12.2 and 12.3. Figure 12.2 shows a Kerr microscope image of magnetic domains in an 8 μm wide stripe of NiFe thin film. The NiFe thin film, labelled "MR stripe" in the figure, is the sensor in a magnetoresistive device. The current leads are used to measure the field-dependent resistivity of the MR stripe. For optimum performance, the magnetic element should remain in a single-domain state. This picture shows a device which was intentionally prepared in a three-domain state by application of an external magnetic field.

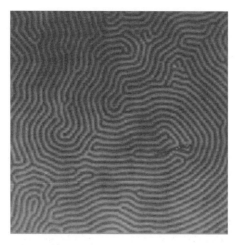

Figure 12.3 Kerr microscope image of domain structure in an yttrium–iron garnet film. Each stripe is ~5 μm wide. Reproduced by permission of Tom Silva, National Institute of Standards and Technology, Boulder, CO.

Figure 12.3 shows a Kerr microscope image of magnetic domains in a film of yttrium–iron garnet (YIG). The magnetization of the film is oriented perpendicular to the film plane. To lower its magnetostatic energy, the film breaks up into this domain pattern which is known as the serpentine domain structure. Each stripe is ~5 μm in width. Kerr microscopy is a powerful means of readily imaging domain patterns in films at the relatively low resolution ~1 μm.

12.1.2 Faraday effect

In the Faraday effect, the plane of polarization of the light beam is rotated as it is *transmitted* through a magnetized sample. In this case the amount of rotation can be several degrees, since the radiation interacts more strongly with the sample than in the Kerr effect. However light is only transmitted for thin samples with low attenuation, and so the Faraday effect can not be used to study bulk samples.

12.1.3 Physical origin of magneto-optic effects

To explain the physics causing Kerr and Faraday rotation, we first need to know that linearly polarized light can be resolved into two oppositely polarized *circular* polarizations:

All photons in circularly polarized beams have the same magnitude of angular momentum (equal to 1) but the angular momentum vector of right circularly polarized light is in the opposite direction to that of left circularly polarized light.

As we discussed in Section 3.3, the magnetization of a magnetic material can cause a Zeeman splitting of the energy levels. For example, if the atomic spin is $\frac{1}{2}$, then each level splits into two levels, with total spin, $S = +\frac{1}{2}$ and $S = -\frac{1}{2}$ respectively.

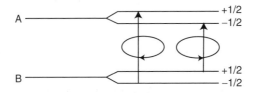

Both energy and angular momentum must be conserved when a photon excites an electron from one of the sub-levels of level B to one of the sub-levels of level A. Therefore, in order to conserve angular momentum, only the following transitions are allowed:

$$S_B = -\tfrac{1}{2} \rightarrow S_A = +\tfrac{1}{2} \text{ with } \Delta L = +1$$
$$S_B = +\tfrac{1}{2} \rightarrow S_A = -\tfrac{1}{2} \text{ with } \Delta L = -1.$$

The photon with $\Delta L = +1$ excites an electron from the $S = -\frac{1}{2}$ state in level B to the $S = +\frac{1}{2}$ state in level A. Similarly the photon with $\Delta L = -1$ excites an electron from the $S = +\frac{1}{2}$ state in level B to the $S = -\frac{1}{2}$ state in level A. So, oppositely polarized photons correspond to different electronic transitions in the atom.

Finally, in level B the electronic population of the two spin states differ one from the other, with the lower-energy state containing statistically more electrons. As a result, the absorption of one of the circular polarizations is greater than that of the other; this phenomenon is known as circular dichroism. When the final circular polarizations are recombined into a linearly polarized beam, the plane of polarization is seen to be rotated compared with that of the incoming beam. The resulting phase difference between the initial and final planes of polarization is called the circular birefringence.

12.2 Magneto-optic recording

Magneto-optic recording combines the advantages of high-density magnetic data storage which were discussed in Chapter 11, with the reduced friction and wear characteristic of conventional optical memories. It also has the additional advantage of being erasable and re-recordable. The principles of magneto-optic recording are illustrated in Fig. 12.4.

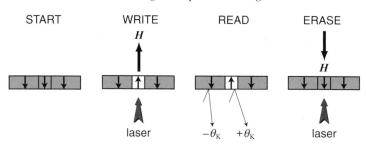

Figure 12.4 The principles of magneto-optic recording.

Before the recording process begins, the magnetization of the entire magnetic film points in the same direction (down say). The area to be written is then heated with a laser to a temperature above the Curie temperature. Then, as the heated area cools, it is magnetized in the opposite direction, either by an applied field, or by the demagnetizing field from the rest of the film. This oppositely magnetized part of the film forms a data bit. The read process uses the Kerr effect with plane-polarized light. If the polarization plane rotates by $+\theta_K$ for upward magnetization, then it must rotate by $-\theta_K$ for downward magnetization. The data can be erased by heating the area with a laser, in the presence of a magnetic field applied in the same direction as the initial magnetization.

The materials requirements on the magneto-optic storage layer are quite stringent. Clearly the magnetic layer must be magneto-optically active so that read-out can be achieved using the Kerr effect. A perpendicular uniaxial magnetic anisotropy constant is desirable since it allows for stable micrometer-sized domains. The Curie temperature should be between 400–600 K – not so high that the laser is unable to heat the material to temperatures above T_C, but not so low that the material is thermally unstable. Both the coercivity and the magnetization should have rather specific temperature dependence. First, H_c needs to be low when zapped with the laser (so that the magnetization can be reversed easily) and high the rest of the time, so that the magnetization doesn't reverse spontaneously. Therefore the $H_c(T)$ curve needs to be steep. The magnetization has the opposite constraints on its temperature dependence – it should be high when zapped with the laser so that there is a large demagnetizing field to reverse the bit, and low the rest of the time so that there is a low demagnetizing field when spontaneous magnetization reversal is undesirable. Additional requirements are a fine-grained or amorphous microstructure and good lateral homogeneity, long-term stability, sensitivity, low media noise and (of course) cheap production.

Good candidate magneto-optic storage media materials are amorphous rare earth–transition metal alloys. Amorphous films are desirable because the noise is low (since there are no grain boundaries) and the films are easily deposited by

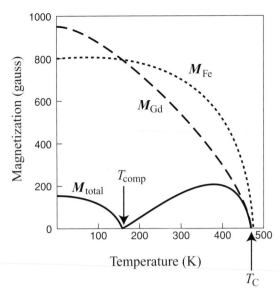

Figure 12.5 Magnetization curve for a Gd–Fe alloy.

sputtering, giving high yields and low cost. Also no post-deposition annealing is required. The rare earth–transition metal alloys are ferrimagnetic, and therefore have a compensation point, as we described in Section 9.1.2.

Typical magnetization versus temperature curves for a representative rare earth–transition metal alloy such as Gd–Fe are shown in Fig. 12.5. Below the compensation temperature, the magnetization of the Gd sublattice is larger than that of the iron sublattice, and so the Gd magnetization lies parallel to the applied field. At the compensation temperature, T_{comp}, the magnetizations of the two sublattices are equal by definition. At higher temperatures the magnetization of the Fe sublattice is largest, and lies parallel to any applied field. Close to T_{comp} the magnetization is small, and the demagnetizing field is small. Also, the coercivity is very large, since the magnetization is zero and so an applied field has no handle to reverse the spin system. As a result the recorded bit is very stable. However heating to just a few degrees above the compensation temperature gives a large reduction in H_c, as shown in the typical H_c versus temperature plot in Fig. 12.6. Therefore the data bits can be recorded easily. Finally, the magneto-optic Kerr rotation comes mainly from the transition metal sublattice, and so it does not show what would be an undesirable singular behavior at the compensation temperature.

An additional advantage of the rare earth–transition metal system is that alloying can be used to tune both T_C and T_{comp} over fairly wide temperature ranges. However the reverse is also true – T_{comp} is strongly dependent on composition, and therefore films need to be uniform if they rely on the properties of the material near the

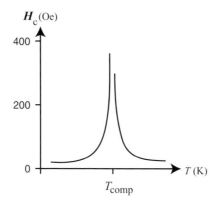

Figure 12.6 Typical variation of coercivity with temperature for a magneto-optic storage medium ferrimagnet. The magnetism is very stable around T_{comp} where the coercive field is high.

compensation temperature for the write process. Another downside is the strong chemical reactivity of the rare earths, particularly in the amorphous phase.

It has been found empirically that *ternary* alloys have a larger Kerr rotation than the simple binary alloys. In particular, TbFeCo is a suitable material because of its large Kerr rotation, large coercivity, and T_C in the range which is appropriate for recording using semiconductor lasers. Unfortunately the Kerr rotation in TbFeCo alloys *decreases* as the wavelength of the laser decreases, and so the material does not perform well at the short wavelengths required for high-density recording. (The diameter of the laser spot decreases as the wavelength decreases allowing smaller and hence more data bits to be written.) Instead NdFeCo has been proposed for future use at shorter wavelength, since its Kerr rotation increases with decreasing wavelength down to around 400 nm. Other potential magneto-optic storage materials are Pt–Co multilayers (which have strong perpendicular anisotropy, high coercivity and high Kerr rotation at blue wavelengths but high T_C and high M_s) and BiFe garnets (which have the largest magneto-optic signal and are chemically very stable, but have a low signal-to-noise ratio and an undesirably high crystallization temperature which limits the choice of substrate). A concise review of these new magneto-optic recording materials can be found in Ref. 61. For a comprehensive discussion, see the book by Gambino and Suzuki, Ref. 62.

12.2.1 Other types of optical storage, and the future of magneto-optic recording

The magneto-optic drive has traditionally been a popular way to back up files on both personal computers and for industrial archiving. The chief assets of MO drives include convenience, modest cost, reliability and removability. The main limitations are that MO drives are slower than hard disk drives and, with recent drops in hard disk drive prices, they can also be more expensive. In addition, the emerging popularity of other optical storage media, including compact discs (CDs) and digital video disks (DVDs) threaten the future of MO storage.

Both CDs and DVDs are safe and reliable media that can provide long-term removable storage for music, data and images. The data bits are structural 'bumps' which are indented cheaply during processing, and no specialized hardware or software is required to read or write the information. The drawback with CDs is their limited storage capacity; a standard CD can store up to around 74 minutes of music. (However, disks can be stored in jukeboxes that can hold 500 CDs at a time.) DVDs are similar to CDs but hold around seven times more data, and are more expensive to produce. The additional storage capacity allows them to store a full-length movie as well as additional information. Since DVDs offer the same storage capacity as MO devices, they are perhaps the most promising near-term technology for reliable data storage.

13

Magnetic semiconductors

" ... quantized spins in quantum dots may prove to be the holy grail for
quantum computing ... "
Stuart A. Wolf "Spintronics; electronics for the next millennium?"
Journal of Superconductivity **13** 195, 2000.

In this chapter we will describe some of the properties and potential uses of three
types of magnetic semiconductors. First, the II–VI diluted magnetic semiconductors
(DMSs), such as (Zn,Mn)Se, which have been studied quite extensively over the
last decade or so. In DMSs some of the cations, which are non-magnetic in conventional semiconductors, are replaced by magnetic transition metal ions. Second, the
fairly new class of III–V diluted magnetic semiconductors, which have generated
great recent excitement following the observation of ferromagnetism in (Ga,Mn)As.
Third, we will discuss some rare earth–group V compounds, particularly ErAs,
which, although not likely to find practical application, show interesting properties
associated with f-electron magnetism.

A large practical motivation for the study of magnetic semiconductors is their
potential for combining semiconducting and magnetic behavior in a single material
system. Such a combination will facilitate the integration of magnetic components
into existing semiconducting processing methods, and also provide compatible
semiconductor–ferromagnet interfaces. As a result, magnetic semiconductors are
viewed as enabling materials for the emerging field of magneto-electronic devices
and technology. Because such devices exploit the fact that the electron has *spin*
as well as charge, they have become known as *spintronic* devices, and their study
is known as *spintronics*. In addition to their potential technological interest, the
study of magnetic semiconductors is revealing a wealth of new and fascinating
physical phenomena, including persistent spin coherence, novel ferromagnetism
and spin-polarized photoluminescence.

This chapter is included in part to introduce a field of magnetic materials that is at the cutting edge of current research. As such, many questions will be left unanswered, and some of the material may seem dated or irrelevant in a few years time. But hopefully we will learn something about how research and technology evolve, and we will have some fun finding out about a fascinating class of materials.

13.1 II–VI diluted magnetic semiconductors – (Zn,Mn)Se

Bulk Mn chalcogenides crystallize in either the hexagonal NiAs structure (α-MnTe) or in the cubic NaCl structure (α-MnSe and α-MnS). Only MnS has been grown in the cubic zincblende structure (β-MnS) in bulk but all Mn chalcogenides can be stabilized artificially in the zincblende structure by epitaxial growth, or by alloying with II–VI semiconductors.[63] The II–VI diluted magnetic semiconductors have been the subject of extensive experimental studies in the past and it has been established that the magnetic structure for sufficiently high manganese concentrations ($x_{Mn} > 0.6$ for MnSe) is characterized by an antiferromagnetic correlation between the Mn^{2+} ($(3d)^5$) magnetic moments. For intermediate manganese concentrations spin-glass structures are found, and for low concentrations ($x_{Mn} < 0.3$ for MnSe) a paramagnetic behavior of the uncorrelated Mn spins becomes dominant.[64] The paramagnetic regime is particularly interesting since the strong sp–d interaction gives rise to a hundred-fold increase of the effective g-factors.[65] The resulting properties include enhanced Zeeman splitting, spin precession and persistent spin coherence, spin-polarized luminescence and spin-polarized transport. We discuss these phenomena in more detail below.

13.1.1 Enhanced Zeeman splitting

When a magnetic field is applied to a semiconductor, the energy of electrons and holes with their spin magnetic moments parallel to the field is lowered, and that of the antiparallel electrons and holes is raised. The difference in energy between the electron–hole pairs of opposite spin polarization is known as the Zeeman splitting. The mechanism of the Zeeman effect in *atoms* was discussed in Section 3.3. In II–VI DMSs, the Mn^{2+} ions become magnetized in the presence of an applied magnetic field. Thus, in addition to the external field, the electrons and holes feel a large magnetization from the Mn^{2+} ions. This results in a Zeeman splitting that can be hundreds of times larger than that in non-magnetic semiconductor quantum structures. This, in turn, results in a giant Faraday rotation, which means that II–VI DMSs have potential application as magneto-optic materials with large magneto-optic coefficients.

13.1.2 Persistent spin coherence

We've seen many times in this book that the spin of an electron is a two-level system whose degeneracy may be split by the application of a magnetic field. If the spins are oriented perpendicular to the magnetic field, and a quantum mechanical wavefunction is constructed which is a superposition of the two energy-split spin states, then the classical magnetization vector precesses about the applied magnetic field as the state evolves in time. While this *Larmor precession* is a classical effect, the underlying mechanism is quantum mechanical, and involves a change in the relative phases of the up-spin and down-spin components of the electron wavefunction. The magnetization vector will continue to precess indefinitely as long as there is no *decoherence* of the quantum mechanical wavefunction.

Similarly, any magnetic ion, such as a Mn^{2+} ion in (Zn,Mn)Se, can be prepared in a state that precesses around an applied magnetic field. In DMSs this can be achieved using circularly polarized light to optically excite spin-polarized excitons. The spin-polarized excitons then couple to the manganese sublattice and transfer their spin polarization.[66] The manganese ions retain their spin polarization, and precess around the applied field, long after the exciton recombination time. Their coherent precession persists for several nanoseconds, even at high temperature, and can be measured using the Faraday rotation technique described in the previous chapter. Typical results are shown in Fig. 13.1.

One of the most promising applications of persistent spin coherence in magnetic semiconductors is in the field of quantum computing and quantum cryptography. Quantum computation is a fundamentally new mode of information processing that can be performed only by harnessing physical phenomena, particularly quantum interference, that are unique to quantum mechanics. In order to build a quantum computer, stable long-lived quantum mechanical states which are coherent are required. The possibility of building a quantum computer in solid state semiconductors is obviously attractive from a practical implementation point of view, particularly if it is operable at room temperature. The details of quantum computing are beyond the scope of this book. A number of good sources now exist for learning more about it; in particular, the text book by Nielsen and Chuang is an excellent reference.[68]

13.1.3 Spin-polarized transport

A characteristic of the transport in two-dimensional electron gases (2DEGs) formed in *non-magnetic* doped semiconductors is the so-called integer quantum Hall effect, in which the application of a magnetic field perpendicular to the 2DEG plane results in a vanishing longitudinal resistance and a quantized Hall resistance. In magnetic 2DEGs, as a result of the enhanced spin-splitting, the energy levels involved in quantum transport are completely spin resolved even at high temperature.[69]

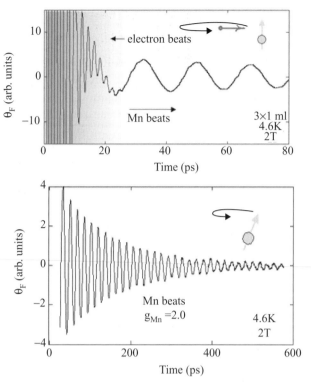

Figure 13.1 Coherent precession of the magnetization in II–VI DMSs measured using Faraday rotation, θ_F. The upper panel shows the last few electron beats decaying to reveal the presence of the manganese beats. The lower panel shows an expanded view of the decay of the Mn^{2+} precession. (3×1 ml = three one-monolayer regions of Mn-doped ZnSe; see text and Figs. 13.2 and 13.3.) From Ref. 67. Copyright (1997) by the American Physical Society.

A magnetoresistance is observed, which is positive at low fields (indicating an increased resistance in the presence of the field), and negative at high fields. The negative high-field magnetoresistance is consistent with the suppression of spin-disorder scattering as the paramagnetic Mn^{2+} ions are aligned in the presence of the field. Research is on-going to provide a detailed model of the magnetoresistance in diluted magnetic semiconductors.

13.1.4 Other architectures

It is also possible (see Fig. 13.2) to grow so-called *digital magnetic heterostructures* (DMHs) of Mn-doped ZnSe using molecular beam epitaxy.[70] In DMHs, the Mn^{2+} ions are constrained to occupy monolayers (or sub-monolayers) within a ZnSe/ZnCdSe quantum well as shown in Fig. 13.3. This arrangement both

Figure 13.2 Schematic of a digital magnetic heterostructure. The gray area represents the ZnCdSe quantum well containing layers of MnSe (black).

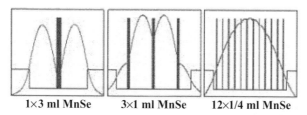

Figure 13.3 Schematic of the conduction band energy profiles and electron wave functions in digital magnetic heterostructures with various distributions of the Mn ions. From Ref. 66. Copyright (1995) by the American Physical Society.

minimizes the tendency of the Mn^{2+} ions to cluster antiferromagnetically, allowing them to respond to an applied magnetic field, and increases the overlap of the electronic wavefunction with the magnetic ions, also shown in Fig. 13.3. Many properties of the II–VI DMHs are superior to those of the corresponding DMSs as a result of this enhanced overlap between the carrier wavefunctions and the localized magnetic moments.

13.2 III–V diluted magnetic semiconductors – (Ga,Mn)As

The III–V diluted magnetic semiconductors are the subject of intense current research interest, both because they are ferromagnetic and because of their compatibility with existing III–V-based technology. The III–V DMSs are obtained by low-temperature molecular beam epitaxy deposition of III–V semiconductors with a transition metal such as Mn. The low-temperature non-equilibrium growth is necessary to prevent the formation of additional phases, and in general only low concentrations (typically around 10^{18} cm^{-3}) of transition metal ions can be incorporated in the non-magnetic matrix. Nevertheless, despite the low concentrations, the systems develop long-range ferromagnetic order with remarkably high Curie temperatures, T_C. For the known III–V-based DMSs the highest Curie temperatures obtained are: $T_C \sim 30$ K for (In,Mn) As,[71] $T_C = 110$ K for (Ga,Mn)As,[71] and, very recently, an unconfirmed report of $T_C = 940$ K for (Ga,Mn)N.[72] More extensive reviews of the properties and potential applications of III–V DMSs can be found in Refs. 71 and 73; here we emphasize a few key points.

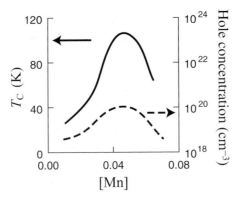

Figure 13.4 Typical variation of Curie temperature (left y axis) and hole concentration (right y axis) with fraction of Mn (x axis) in (Ga,Mn)As.

Three important features underlie the ferromagnetic order of (Ga,Mn)As and other Mn-doped III–V DMSs: (i) Mn^{2+} ions substitute for the Ga^{3+} cations in the zincblende lattice providing localized magnetic moments ($S = 5/2$); (ii) there are free holes in the system, although the actual concentration is much smaller than the Mn concentration (despite the fact that the nominal valence suggests that the two concentrations should be identical); and (iii) the hole spins couple antiferromagnetically with the Mn spins, due to a dynamic p–d coupling. Curie temperatures have been shown to correlate with the hole concentration – a schematic of typical data is shown in Fig. 13.4.

A number of models have been proposed to describe the ferromagnetism in III–V DMSs, the most widely accepted being the Zener model,[74] in which antiferromagnetic exchange-bias coupling partially spin-polarizes the itinerant holes, which in turn cause an alignment of the localized Mn moments. Within the Zener model, the interaction Hamiltonian between the hole spin $\vec{s}$ and the Mn spin $\vec{S}$ is

$$H = -N_0 \beta \vec{s} \cdot \vec{S}, \tag{13.1}$$

where N_0 is the concentration of the cation sites and β is the p–d exchange integral. The product $N_0\beta$ is usually called the exchange constant. If one simply uses the mean-field approximation,[74, 75] in which the magnetizations of both carriers and Mn ions are considered to be uniform in space, one obtains an expression for the Curie temperature

$$T_C = \frac{x N_0 S(S+1)\beta^2 \chi_s}{3k_B(g^*\mu_B)^2}, \tag{13.2}$$

where χ_s is the magnetic susceptibility of the free carriers (holes in this case), g^* is their g-factor, k_B is the Boltzmann constant and μ_B the Bohr magneton. This expression gives T_Cs in reasonable agreement with measured transition temperatures, and can be greatly refined by including a detailed description of the band structure of the underlying non-magnetic semiconductors or by going beyond the mean-field approximation to incorporate correlation effects.

However the ferromagnetism of (Ga,Mn)As is very sensitive to sample history, including the growth conditions[71] and post-growth processing.[76, 77] Since the growth dynamics certainly affects the microscopic configuration of the samples, this suggests that knowledge of the local chemical environment is crucial for understanding and modeling the properties correctly. First-principles density functional calculations have been invaluable in elucidating the detailed effects of microscopic configuration on magnetic properties, including the influence of As antisites on the ferromagnetic Curie temperature,[78] and the role of the arrangement of Mn ions on transport.[79]

Clearly, for device applications it is desirable to find a material with a Curie temperature at or above room temperature. In addition to the current research aimed at understanding the origin of the ferromagnetism in (Ga,Mn)As, there is also considerable exploration of related materials in the search for higher Curie temperatures. In particular, ferromagnetism with T_C above room temperature has been reported in several other DMSs, including $Cd_{1-x}Mn_xGeP_2$,[80] $Ti_{1-x}Co_xO_2$,[81] and $Zn_{1-x}Co_xO$.[82]

13.3 Rare earth–group V compounds – ErAs

Finally we mention the rare earth–group V compounds, which are attractive because they are thermodynamically stable, and are closely lattice-matched to III–V semiconductors, allowing study of their fundamental properties. However typically their antiferromagnetic ordering persists only to a few kelvin, therefore they are not likely to find technological application for their magnetic properties!

ErAs is one of the most well-studied rare earth–group V compounds. It crystallizes in the rock-salt structure and is a semimetal. The band structure, calculated using the linear-muffin-tin-orbital (LMTO) method is shown in Fig. 13.5.[83] ErAs orders antiferromagnetically with a Néel temperature of 4 K. It grows readily on GaAs, and can be alloyed with Sc to improve the lattice match. However it is difficult to re-grow GaAs onto ErAs, because of the different crystal symmetry and the wetting behavior of the materials. A practical solution for achieving multilayers is to grow *islands* rather than films of ErAs.[84]

The resulting materials have interesting magnetotransport properties, with a cusp in the magnetoresistance at around 1 T (sketched in Fig. 13.6) that has been attributed to spin-disorder scattering. Also, the experimental phase diagram, sketched in Fig. 13.7, indicates a field-induced antiferromagnetic (AFM) to paramagnetic (PM) phase transition.

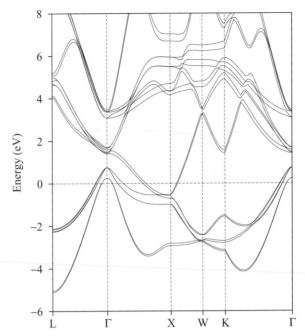

Figure 13.5 Band structure of ErAs, calculated along the high-symmetry axes of the cubic Brillouin zone. The spin–orbit coupling is included. From Ref. 83. Copyright (1997) by the American Physical Society.

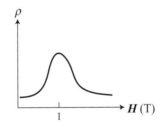

Figure 13.6 Typical magnetoresistance of ErAs–GaAs films.

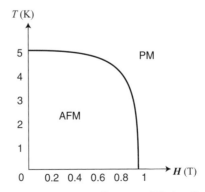

Figure 13.7 Typical phase diagram of ErAs–GaAs films.

13.4 Summary

In this chapter we have introduced some of the magnetic semiconductor materials that are of great current research interest. It is likely that some of these materials will result in new technological applications, both by facilitating improvements of existing device architectures and by introducing new storage and processing paradigms. The ferromagnetic semiconductors are particularly promising since they have compatible interfaces with conventional semiconductors (allowing injection of spin-polarized electrons and holes) and can be integrated using existing semiconductor processing techniques. Even in the unlikely event that no relevant applications are found, research in this field has revealed a wealth of new fundamental physics and will undoubtedly continue to do so for years to come.

Epilogue

A paramagnetic poem
by Lucy Popova. Reproduced with permission

You're telling me
That frogs explode at eight
That they can levitate
But that's too little, too late
   We have already started
Cooling.
It's all paramagnetic
I am no paramedic
  But I feel
    How the pulse
     Between us
Is slowing down.

There is no word
Good enough to disagree
Now, like hydrogen at three
We are frozen. We are free
   I've learned that
Cooling
Can be paramagnetic
It's all para-pathetic
  But I see
    That the space
     Between us
Is light years wide.

I've read in books
So many reasons why
The frogs would never fly
But you will make one try
   And then you'll see that
Cooling
Was all paramagnetic
And when there is no static
  I will hear
    How you call
     Through the cold
To say "hello".

Solutions to exercises

Chapter 1

1.1 It's easier to use the Biot–Savart law to calculate the field at the center of a circular coil of current.

Divide the coil into elements of arc length δl, each of which contributes a field

$$\delta H = \frac{1}{4\pi a^2} I \delta l \times \hat{u} \tag{S.1}$$

at the center of the coil, as shown in Fig. S.1.

Then sum over all the elements to get the *total* field:

$$H = \sum \frac{1}{4\pi a^2} I \delta l \sin 90°. \tag{S.2}$$

But $\sum \delta l = 2\pi a$ (the circumference of the coil) and $\sin 90° = 1$, so

$$H = \frac{I}{2a}. \tag{S.3}$$

The SI units of H are A/m.

1.2a We'll use the Biot–Savart law again, this time to derive the field on the *axis* of a circular coil. The geometry of the problem is shown in Fig. S.2.

Each element δl contributes a field δH at a distance r from the element, where

$$\delta H = \frac{1}{4\pi r^2} I \delta l \times \hat{u}$$

$$= \frac{1}{4\pi r^2} I \delta l \sin 90°$$

$$= \frac{1}{4\pi r^2} I \delta l. \tag{S.4}$$

By symmetry, $\delta H_{\text{tangential}} = 0$, and $\delta H_{\text{axial}} = \delta H \sin \alpha$. So

$$\frac{\delta H_{\text{axial}}}{\sin \alpha} = \frac{1}{4\pi r^2} I \delta l. \tag{S.5}$$

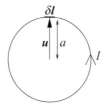

Figure S.1 Using the Biot–Savart law to derive the field at the center of a circular coil.

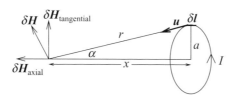

Figure S.2 Geometry for derivation of the field on the axis of a circular coil.

But $r = a/\sin \alpha$, giving

$$\delta H_{\text{axial}} = \frac{1}{4\pi a^2} I \sin^3 \alpha \delta l. \tag{S.6}$$

Integrating around the coil, $\int \delta l = 2\pi a$, so

$$H_{\text{axial}} = \frac{I}{2a} \sin^3 \alpha \tag{S.7}$$

$$= \frac{I a^2}{2(a^2 + x^2)^{3/2}}. \tag{S.8}$$

1.2b   For a general, off-axis point, the Biot–Savart law can still be used to obtain the magnetic field contribution, δH, from a current element $I \delta l$ at a distance r from the coil:

$$\delta H = \frac{1}{4\pi a^2} I \delta l \times \hat{u} \tag{S.9}$$

$$= \frac{I \delta l \sin \theta}{4\pi r^2}. \tag{S.10}$$

Here r is a function of θ, and H can be obtained numerically by an elliptic integral. Since a knowledge of magnetic fields is very important in device design, many sophisticated numerical techniques have been developed for their calculation in cases of general symmetry. C.W. Trowbridge, *IEEE Trans. Mag.* **24**, 13 (1988) is a good review.

1.3a   We use the expression which we derived in Solution 1.2a above,

$$H_{\text{axial}} = \frac{I a^2}{2(a^2 + x^2)^{3/2}} \tag{S.11}$$

with $a = 1$ Å $= 10^{-10}$ m and $x = 3$ Å $= 3 \times 10^{-10}$ m.

To calculate the current I, we use the fact that the angular momentum (which in general is given by $m_e va$) is $\hbar$ J s. So

$$v = \frac{\hbar}{m_e a} \frac{\text{J s}}{\text{kg m}} = \frac{\hbar}{m_e a} \frac{\text{m}}{\text{s}} \tag{S.12}$$

and the current,

$$
\begin{aligned}
I &= \frac{\text{charge}}{\text{time}} \\
&= \frac{e}{\text{distance/speed}} \\
&= e \frac{v}{2\pi a} \\
&= \frac{e}{2\pi a} \frac{\hbar}{m_e a} \\
&= 2.952 \times 10^{-4} \text{ A}. \tag{S.13}
\end{aligned}
$$

Then

$$H = \frac{2.952 \times 10^{-4} \times (10^{-10})^2}{2[(10^{-10})^2 + (3 \times 10^{-10})^2]^{3/2}} \frac{\text{A m}^2}{\text{m}^3}$$
$$= 46\,675.7\,\text{A/m} = 586\,\text{Oe}. \tag{S.14}$$

1.3b   The magnetic dipole moment, $\boldsymbol{m}$, is given by

$$
\begin{aligned}
\boldsymbol{m} &= I\,A \\
&= \frac{ev}{2\pi a} \pi a^2 \\
&= \frac{eva}{2} \\
&= \frac{a}{2} \frac{e\hbar}{m_e a} \\
&= \frac{e\hbar}{2m_e} \\
&= 9.274 \times 10^{-24} \text{ A m}^2 \text{ or J/T}. \tag{S.15}
\end{aligned}
$$

This number is the Bohr magneton, μ_B, and is the natural unit of magnetic moment. In cgs units, the Bohr magneton is equal to $e\hbar/2m_e c = 0.927 \times 10^{-20}$ erg/Oe. (Remember that the value of the speed of light, c, in cgs units is 4.80×10^{-10} esu, and the value of h is 6.62×10^{-27} erg s.)

1.3c   The magnetic dipolar energy,

$$E = -\mu_0 \boldsymbol{m} \cdot \boldsymbol{H}$$
$$= 1.256 \times 10^{-6} \text{ weber/A m} \times -9.274 \times 10^{-24} \text{ A m}^2 \times 46\,675.7 \text{ A/m} \tag{S.16}$$
$$= -5.44 \times 10^{-25} \text{ J}. \tag{S.17}$$

Note that this number is very small, so it is unlikely that the parallel alignment of magnetic dipole moments in ferromagnetic materials results from a magnetic dipolar interaction.

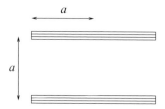

Figure S.3 Derivation of the field on the axis of a circular coil.

1.4 The geometry of the problem is shown in Fig. S.3. These are known as Helmholtz coils.

In Solution 1.2a we derived the expression for the field produced by a current flowing in a circular coil of radius a, at a distance x from the coil along its axis, and obtained

$$H = \frac{Ia^2}{2(a^2 + x^2)^{3/2}} = \frac{I}{2a}\left(1 + \frac{x^2}{a^2}\right)^{-3/2}. \tag{S.18}$$

In this case, if there are N turns of wire forming each coil, the total effective current is NI.

(a) If the coils are wound in the same direction, then the fields produced by each coil add to each other, so

$$H = \frac{NI}{2a}\left(1 + \frac{x^2}{a^2}\right)^{-3/2} + \frac{NI}{2a}\left(1 + \frac{(a-x)^2}{a^2}\right)^{-3/2}. \tag{S.19}$$

If $a = 1$, then the values of the field for a range of x values are as given in the following table

x	H
0.25	$NI/2\,(1.0625^{-3/2} + 1.5625^{-3/2}) = 1.43NI/2$
0.5	$NI/2\,(1.25^{-3/2} + 1.25^{-3/2}) \quad\;\; = 1.43NI/2$
0.75	$NI/2\,(1.5625^{-3/2} + 1.0625^{-3/2}) = 1.43NI/2$

That is, the field between two Helmholtz coils wound in the same direction is constant. As a consequence they are used whenever a constant magnetic field strength is needed over a large volume of space. However they are restricted to low-field applications, because the field produced is much lower than that produced by a solenoid carrying the same current flow.

(b) If the coils are wound in the opposite direction, then the fields produced by each coil subtract from each other, so

$$H = \frac{NI}{2a}\left[\left(1 + \frac{x^2}{a^2}\right)^{-3/2} - \left(1 + \frac{(a-x)^2}{a^2}\right)^{-3/2}\right]. \tag{S.20}$$

The field gradient, dH/dx is

$$\frac{dH}{dx} = \frac{-3NI}{2a}\left[x\left(1 + \frac{x^2}{a^2}\right)^{-5/2} + (a-x)\left(1 + \frac{(a-x)^2}{a^2}\right)^{-5/2}\right]. \quad \text{(S.21)}$$

The numerical values of field and field gradient with $a = 1$ are as given in the following table:

x	$H(\times NI/2)$	$dH/dx(\times -3NI/2)$
0.25	$1.0625^{-3/2} - 1.5625^{-3/2} = 0.40$	$0.25 \times 1.0625^{-5/2} + 0.75 \times 1.5625^{-5/2} = 0.46$
0.5	$1.25^{-3/2} - 1.25^{-3/2} \quad\quad = 0$	$0.5 \times 1.25^{-5/2} + 0.5 \times 1.25^{-5/2} \quad\quad = 0.57$
0.75	$1.5625^{-3/2} - 1.0625^{-3/2} = -0.40$	$0.75 \times 1.5625^{-5/2} + 0.25 \times 1.0625^{-5/2} = 0.46$

OK, so I didn't choose good numbers for this question. I wanted to illustrate that Helmholtz coils wound in opposite directions give rise to a constant field gradient. In fact, if we had chosen x values nearer to the center of the coil, we would have found that the field gradient was approximately constant. Helmholtz coils wound in opposite directions are used whenever a constant field gradient is required, for example to exert a constant force.

Chapter 2

2.1(a)   1 erg $= 10^{-7}$ J, and 1 Oe $= 1/4\pi \times 10^{-3}$ A/m $= 10^{-4}$ T, so

$$10\,000 \text{ erg/Oe} = 10\,000 \times 10^{-7}\,\text{J/Oe} = \frac{10\,000 \times 10^{-7}}{10^{-4}}\,\text{J/T} = 10 \text{ J/T}. \quad \text{(S.22)}$$

(b) 1 in $= 2.54$ cm $= 2.54 \times 10^{-2}$ m. Therefore the volume of the cylinder, which is equal to $\pi r^2 l$, where r is the radius and l is the length, is equal to 128.704 cm^3, which is $1.287\,04 \times 10^{-4}$ m^3.

The magnetization, M, is defined to be the magnetic moment per unit volume. In cgs units,

$$M = \frac{m}{V} = \frac{10\,000 \text{ erg/Oe}}{128.704 \text{ cm}^3} = 77.70 \text{ emu/cm}^3 \quad \text{(S.23)}$$

since 1 erg/Oe $= 1$ emu. In SI units,

$$M = \frac{m}{V} = \frac{10 \text{ J T}^{-1}}{1.287\,04 \times 10^{-4}\,\text{m}^3} = 77.7 \times 10^3\,\frac{\text{kg m}^2\,\text{s}^{-2}}{\text{m}^3\,\text{kg s}^{-2}\text{A}^{-1}} = 77.7 \times 10^3\,\text{A/m}. \quad \text{(S.24)}$$

(c) For a current loop, the magnetic moment, $m = IA$. For a solenoid with N loops, the magnetic moment is NIA. Working in SI units,

$$10 \text{ J/T} = 100 \times I \times \pi \times (0.0127)^2\,\text{m}^2, \quad \text{(S.25)}$$

therefore

$$I = 197.3 \text{ J/T m}^2 = 197.3 \text{ A}. \tag{S.26}$$

Chapter 3

3.1 The magnitude of the total magnetic moment of an atom is equal to $g\sqrt{J(J+1)}\mu_B$, and the component of that moment projected along the field direction is $-gM_J\mu_B$.

When $J = 1$, $\sqrt{J(J+1)} = \sqrt{2}$ and $M_J = -1, 0$ or 1. Therefore for $g = 2$ the total moment is $2\sqrt{2}\mu_B$, and the component of the magnetic moment along the field direction can be $-2\mu_B$, 0 or $+2\mu_B$. Note that in all cases the component along the field direction is *less* than the total magnetic moment.

3.2a The electronic configuration of an Fe atom is

$$(1s)^2(2s)^2(2p)^6(3s)^2(3p)^6(4s)^2(3d)^6. \tag{S.27}$$

Therefore, because the transition elements give up their 4s electrons before their 3d electrons on ionization, the electronic configuration of an Fe^{2+} *ion* is

$$(1s)^2(2s)^2(2p)^6(3s)^2(3p)^6(3d)^6. \tag{S.28}$$

3.2b Hund's first rule tells us that the electrons maximize their total spin, S. Therefore they arrange themselves one electron per d orbital with parallel spins before pairing up with opposite spins in the same orbital. For iron, the resulting configuration looks like this:

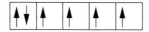

Therefore the total spin, $S = 4 \times \frac{1}{2} = 2$.

The five d orbitals have m_l values of $-2, -1, 0, 1$ and 2. Depending on which d orbital contains two electrons, the total M_L can be $-2, -1, 0, 1$ or 2. Therefore, since $M_L = -L$, $-L+1, \ldots, 0, \ldots, L-1, L$, the total orbital quantum number, L, must be equal to 2.

Finally, from Hund's third rule, because the shell is more than half full, $J = L + S = 4$.

3.2c The Landé g-factor,

$$\begin{aligned} g &= 1 + \frac{J(J+1) + S(S+1) - L(L+1)}{2J(J+1)} \\ &= 1 + \frac{20 + 6 - 6}{40} \\ &= 1.5 \end{aligned} \tag{S.29}$$

Note that, since $S = 2$ and $L = 2$, the g value is exactly halfway between that for the $S = 0$ case ($g = 1$) and the spin-only case ($g = 2$).

3.2d The *total* magnetic moment, $g\sqrt{J(J+1)}\mu_B = 1.5 \times \sqrt{4 \times 5} = 6.7\mu_B$. Since $J = 4$, $M_J = -4, -3, -2, -1, 0, 1, 2, 3$ or 4. Therefore the components of the magnetic moment along the field direction, $-gM_J\mu_B$, can take the values $6\mu_B$, $4.5\mu_B$, $3\mu_B$, $1.5\mu_B$, 0, $-1.5\mu_B$, $-3\mu_B$, $-4.5\mu_B$ or $-6\mu_B$.

In Solution 1.3b we calculated the magnetic moment of a 'classical' orbiting electron and obtained a result of $1\mu_B$. This is the same order of magnitude as the results obtained here.

3.2e If L were equal to zero, then $J = S = 2$ and $g = 2$. So the total magnetic moment would be $2\sqrt{6}\mu_B = 4.9\mu_B$. This is in good agreement with the measured value of $5.4\mu_B$, whereas the moment we calculated using the *total* angular momentum ($6.7\mu_B$) does not agree well with the experimental value. This is a manifestation of a phenomenon is known as *quenching* of the orbital angular momentum that we discuss in Section 5.3.

Chapter 4

4.1 The expression for the diamagnetic susceptibility in SI units is

$$\chi = -\frac{N\mu_0 Z e^2}{6m_e}\langle r^2\rangle_{av}. \tag{S.30}$$

Here N is the number of atoms per unit volume, ($= N_A\rho/A$, where N_A is Avogadro's number (the number of atoms per mole), ρ is the density and A is the atomic weight), μ_0 is the permeability of free space, Z is the number of electrons per atom, e is the electronic charge, m_e is the mass of the electron, and $\langle r^2\rangle_{av}$ is the average distance squared of the electrons from the nucleus.

For carbon, $Z = 6$ and $A = 12\,\text{g/mol}$, so

$$\begin{aligned}
\chi &= -\frac{N\mu_0 Z e^2}{6m_e}\langle r^2\rangle_{av} \\
&= -\frac{6.022\times10^{23}\,\text{mol}^{-1}\times 2220\,\text{kg m}^{-3}}{12\times10^{-3}\,\text{kg mol}^{-1}} \\
&\quad\times\frac{1.256\times10^{-6}\,\text{H m}^{-1}\times 6(1.60\times10^{-19})^2\,\text{C}^2\times(0.7\times10^{-10})^2\,\text{m}^2}{6\times9.109\times10^{-31}\,\text{kg}} \\
&= -19.33\times10^{-6}\,\text{H C}^2\,\text{m}^{-2}\,\text{kg}^{-1} \\
&= -19.33\times10^{-6}. \tag{S.31}
\end{aligned}$$

This is reasonably close to the experimental value of -13.82×10^{-6}.

In cgs units the corresponding expression for the susceptibility is

$$\begin{aligned}
\chi &= -\frac{N Z e^2}{6m_e c^2}\langle r^2\rangle_{av} \\
&= -\frac{6.022\times10^{23}\,\text{mol}^{-1}\times 2.22\,\text{g cm}^{-3}}{12\,\text{g mol}^{-1}} \\
&\quad\times\frac{6\times(4.8\times10^{-10})^2\,\text{esu}^2\times(0.7\times10^{-8})^2\,\text{cm}^2}{6\times9.109\times10^{-28}\,\text{g}\times(3\times10^{10})^2\,\text{cm s}^{-2}} \\
&= -1.5\times10^{-6}\,\text{emu/cm}^3\,\text{Oe}. \tag{S.32}
\end{aligned}$$

For most materials, the value calculated using the classical Langevin model only shows order-of-magnitude agreement with experiment. Possible sources of the discrepancies between theory and experiment include the following.

- The application of Lenz's law (which was determined for electrical circuits) on the atomic scale.

- Difficulty in calculating or measuring $\langle r^2 \rangle_{\text{av}}$. In particular, χ depends on the choice of origin for computing $\langle r^2 \rangle_{\text{av}}$.

- We've assumed that the electrons are orbiting their nucleus. Therefore, we expect a poor description of itinerant conduction electrons.

- We've assumed that the system is spherically symmetric.

- We might expect that the use of classical mechanics should introduce errors. However, a full quantum mechanical derivation actually gives the same result.

Chapter 5

5.1 The Brillouin function, $B_J(\alpha)$, is given by

$$B_J(\alpha) = \frac{2J+1}{2J} \coth\left(\frac{2J+1}{2J}\alpha\right) - \frac{1}{2J} \coth\left(\frac{\alpha}{2J}\right). \tag{S.33}$$

As $J \to \infty$, $2J + 1 \to 2J$, and so $(2J + 1)/2J \to 1$. Therefore the first term tends to $\coth \alpha$. The second term tends to the coth of a very small number, so we can use the series expansion

$$\coth(x) = \frac{1}{x} + \frac{x}{3} - \frac{x^3}{45} + \cdots \tag{S.34}$$

which is valid for small x. The second term then becomes

$$-\frac{1}{2J} \coth\left(\frac{\alpha}{2J}\right) = -\frac{1}{2J}\frac{2J}{\alpha} - \frac{1}{2J}\frac{\alpha}{6J} + \frac{1}{2J}\frac{1}{45}\frac{\alpha^3}{(2J)^3} - \cdots \to -\frac{1}{\alpha} \text{ as } J \to \infty. \tag{S.35}$$

Therefore

$$B_J(\alpha) \to \coth(\alpha) - \frac{1}{\alpha} = L(\alpha) \text{ as } J \to \infty. \tag{S.36}$$

As $J \to \frac{1}{2}$, $(2J + 1)/2J \to 2$, and $2J \to 1$. So at $J = \frac{1}{2}$,

$$B_J(\alpha) = 2\coth(2\alpha) - \coth(\alpha)$$

$$= 2\frac{e^{2\alpha} + e^{-2\alpha}}{e^{2\alpha} - e^{-2\alpha}} - \frac{e^{\alpha} + e^{-\alpha}}{e^{\alpha} - e^{-\alpha}}$$

$$= \frac{2e^{2\alpha} + 2e^{-2\alpha} - (e^{\alpha} + e^{-\alpha})^2}{(e^{\alpha} + e^{-\alpha})(e^{\alpha} - e^{-\alpha})}$$

$$= \frac{e^{2\alpha} + e^{-2\alpha} - 2}{(e^{\alpha} + e^{-\alpha})(e^{\alpha} - e^{-\alpha})}$$

$$= \frac{(e^{\alpha} - e^{-\alpha})(e^{\alpha} - e^{-\alpha})}{(e^{\alpha} + e^{-\alpha})(e^{\alpha} - e^{-\alpha})}$$

$$= \tanh(\alpha). \tag{S.37}$$

As $\alpha \to 0$, $\coth[(2J + 1)\alpha/2J] \to 2J/(2J + 1)\alpha + (2J + 1)\alpha/3 \times 2J$ and $\coth(\alpha/2J)$
$\to 2J/\alpha + \alpha/3 \times 2J$. So

$$B_J(\alpha) \to \frac{2J + 1}{2J} \times \frac{2J}{\alpha(2J + 1)} - \frac{1}{2J} \times \frac{2J}{\alpha} + \left(\frac{2J + 1}{2J}\right)^2 \frac{\alpha}{3} - \left(\frac{1}{2J}\right)^2 \frac{\alpha}{3}$$

$$= \frac{[(2J + 1)^2 - 1]\alpha}{12J^2}$$

$$= \alpha \frac{(J + 1)}{3J}. \tag{S.38}$$

5.2   Let's work in SI units and use the quantum mechanical form for the paramagnetic susceptibility within the Langevin localized-moment model. Then

$$\chi = \frac{Ng^2 J(J + 1)\mu_0\mu_B^2}{3k_B T}. \tag{S.39}$$

Substituting the values $J = 1$, $g = 2$, $\mu_0 = 4\pi \times 10^{-7}$ H/m, $\mu_B = 9.274 \times 10^{-24}$ J/T, $k_B = 1.380\,662 \times 10^{-23}$ J/K and $T = 273$ K gives

$$\chi = \frac{N \times 8 \times 4\pi \times 10^{-7} \times (9.274 \times 10^{-24})^2}{3 \times 1.380\,662 \times 10^{-23} \times 273} \frac{\text{H m}^{-1} \text{J}^2 \text{T}^{-2}}{\text{J K}^{-1} \text{K}}$$

$$= 7.6465 \times 10^{-32} N \frac{\text{H J}}{\text{m T}^2}$$

$$= 7.6465 \times 10^{-32} N \text{ m}^3. \tag{S.40}$$

Since the SI susceptibility should be dimensionless we need to obtain N as a number per m^3. We'll use the ideal gas law to do that. Using $PV = nRT$, where n is the number of moles of atoms with $P = 1$ atm $= 101\,325$ N m^{-2}, $R = 8.314\,41$ J mol^{-1} K^{-1}, $T = 273$ K and the volume, $V = 1$ m^3 gives the number of atoms per m^3:

$$N = \frac{PV \times N_A}{RT}$$

$$= \frac{101\,325 \times 1 \times 6.022 \times 10^{23}}{8.314\,41 \times 273} \frac{\text{N m}^{-2} \text{m}^3 \text{mol}^{-1}}{\text{J mol}^{-1} \text{K}^{-1} \text{K}}$$

$$= 2.688 \times 10^{25}. \tag{S.41}$$

Substituting in Eqn S.40 gives

$$\chi = 2.056 \times 10^{-6}. \tag{S.42}$$

Note that this is a small and positive number.

5.3(a)   Remember that the magnitude of the total magnetic moment of a spin S is equal to $g_e\mu_B\sqrt{S(S + 1)}$ and the component along a specific direction is given by $-g_e\mu_B m_s$. Here g_e is the g-factor of the electron, which is equal to 2, m_s can take values of $\frac{1}{2}$ and $-\frac{1}{2}$ and μ_B is the Bohr magneton. So the total magnetic moment when $J = 1$ and $g = 2$ is $\sqrt{3}\mu_B$ and the allowed values along the z axis are $\pm\mu_B$.

(b) Since the magnetic energy, $E = -\mathbf{m} \cdot \mathbf{H}$, the allowed magnetic energies are $\mp \mu_B H$, in an applied field $\mathbf{H}$, of magnitude H.

(c) In this case the partition function, $Z = \Sigma_i e^{-E_i/k_B T} = e^{\mu_B H/k_B T} + e^{-\mu_B H/k_B T} = 2\cosh(\mu_B H/k_B T)$. So the average magnetization per spin,

$$\langle \mathbf{M} \rangle = \frac{1}{Z} \Sigma_i \mathbf{m}_i e^{-E_i/k_B T}$$

$$= \frac{\mu_B}{Z} (e^{\mu_B H/k_B T} - e^{-\mu_B H/k_B T})$$

$$= \mu_B \tanh \left(\frac{\mu_B H}{k_B T} \right).$$

So the total magnetization, M is given by

$$M = n\mu_B \tanh \left(\frac{\mu_B H}{k_B T} \right)$$

where n is the number of spins per unit volume.

(d) For a given field, the magnetization decreases from the zero-temperature value of $n\mu_B$ to the high-temperature value of zero as the temperature T increases from zero to ∞. For $n = 3.7 \times 10^{28}$ m^{-3}, the zero-temperature saturation magnetization is

$$M_s = 3.7 \times 10^{28} \text{ m}^{-3} \times 9.274 \times 10^{-24} \text{ J T}^{-1}$$
$$= 3.43 \times 10^5 \text{ A/m}.$$

At zero temperature the spins are perfectly aligned by an external field because there is no thermal energy available to randomize the spin directions (and thus increase the *entropy*). At infinitely high temperature there is enough thermal energy to randomize the spin directions (giving a net magnetization of zero) even in the presence of an external magnetic field.

(e) As $x \to 0$, $\tanh(x) \to x$, therefore as $H \to 0$, $\tanh(\mu_B H/k_B T) \to \mu_B H/k_B T$. So the magnetization, $M \to n\mu_B^2 H/k_B T$.

The susceptibility,

$$\chi = \frac{M}{H}$$

$$= \frac{n\mu_B^2}{k_B} \frac{1}{T},$$

i.e. the susceptibility is inversely proportional to temperature, and diverges only as $T \to 0$. Note that this is Curie's law.

At room temperature, when $T = 300$ K,

$$\chi = \frac{3.7 \times 10^{28} \text{ m}^{-3} \times (9.274 \times 10^{-24})^2 \text{ J}^2 \text{ T}^{-2}}{1.381 \times 10^{-23} \text{ J K}^{-1} \times 300 \text{ K}}$$

$$= 768.11 \frac{\text{J}^3 \text{ T}^{-2}}{\text{m}^3}$$

or, multiplying by μ_0 to convert to dimensionless units, $\chi = 0.000\,965$.

(f) The behavior described by this non-interacting spin system is *paramagnetic*. The system obeys Curie-law behavior, and there is no phase transition to a magnetically ordered state. In order to describe ferromagnetic behavior we would have to add interactions to our model. The interactions would have to be such that the energy of the system was lower when neighboring spins were aligned parallel to each other, compared to when they were not aligned at all, or had some other (for example antiparallel) alignment.

Chapter 6

6.1 Equating, at the origin, the slope of the magnetization described by the Langevin function (which is $\frac{1}{3} \times Nm$), with the slope of the straight line representing magnetization by the molecular field, gives

$$\frac{k_B T_C}{m\gamma} = \frac{1}{3} \times Nm. \tag{S.43}$$

So, if the Curie temperature is known, then the molecular field constant can be extracted:

$$\gamma = \frac{3k_B T_C}{Nm^2}. \tag{S.44}$$

Similarly the Weiss molecular field, $H_W = \gamma M = \gamma Nm = 3k_B T_C/m$. For Ni, the magnetic moment per atom, $m = 0.6\mu_B$, and the Curie temperature, $T_C = 628.3$ K. Therefore

$$H_W = \frac{3k_B T_C}{m} = \frac{3 \times 1.380\,662 \times 10^{-23}\,\text{J K}^{-1} \times 628.3\,\text{K}}{0.6 \times 9.274 \times 10^{-24}\,\text{J T}^{-1}} = 4676.89\,\text{T}. \tag{S.45}$$

This is a *very* large field!

6.2(a) In Chapter 1, we calculated that the field generated by such an electron was 46 675.7 A/m, and that the magnetic moment was μ_B. Assuming a 'classical' electron, so that $T_C = Nm^2\gamma/3k_B$, and taking $\gamma = H/M = H/Nm$. Then

$$T_C = \frac{mH}{3k_B}$$

$$= \frac{9.274 \times 10^{-24}\,\text{J T}^{-1} \times 46\,675.7\,\text{A m}^{-1}}{3 \times 1.380\,662 \times 10^{-23}\,\text{J K}^{-1}}$$

$$= 10\,450.794/\text{K m H}^{-1}$$

$$= 0.0131\,\text{K} \tag{S.46}$$

(multiplying by $\mu_0 = 1.25 \times 10^{-6}$ H m^{-1}, to convert to kelvin.) Note that this is a very small number!

(b) In a field of 50 Oe the magnetic dipole energy would be

$$E = -\mu_0 m \cdot H$$

$$= -9.274 \times 10^{-24}\,\text{J T}^{-1} \times \left(50 \times \frac{1000}{4\pi}\right)\,\text{A m}^{-1} \times 1.25 \times 10^{-6}\,\text{H m}^{-1}$$

$$= -4.637 \times 10^{-26}\,\text{J}. \tag{S.47}$$

At 298 K the thermal energy, $k_B T = 4.11 \times 10^{-21}$ J, which is *five orders of magnitude larger* than the magnetic energy! Therefore a field of around 50 Oe would have no effect on aligning electronic magnetic moments at room temperature. We can conclude that the effective internal 'field' which aligns the magnetic moments of ferromagnets spontaneously is much larger than 50 Oe.

6.3   Review question

(a) For this problem it's much easier in this case to use the Biot–Savart law, as we did in Solution 1.2a. This gives us the following expression for the magnetic field at a distance x away from a current-carrying circular loop, on the axis of the loop:

$$H = \frac{I}{2a} \sin^3 \alpha \tag{S.48}$$

$$= \frac{Ia^2}{2(a^2 + x^2)^{3/2}}. \tag{S.49}$$

To estimate the field generated by a Ni atom at its neighbor in a solid, let's pretend that the electrons in the Ni atom are orbiting around the nucleus with a radius $a = 1$ Å, and that the neighboring Ni atom is a distance $x = 3$ Å from the first Ni atom.

To estimate the current I, we use the fact that the angular momentum of an electron (which in general is given by $m_e v a$) is of the order of $\hbar$ J s. So

$$v = \frac{\hbar}{m_e a} \frac{\text{J s}}{\text{kg m}} = \frac{\hbar}{m_e a} \frac{\text{m}}{\text{s}} \tag{S.50}$$

and the current,

$$I = \frac{\text{charge}}{\text{time}}$$

$$= \frac{e}{\text{distance/speed}}$$

$$= e \frac{v}{2\pi a}$$

$$= \frac{e}{2\pi a} \frac{\hbar}{m_e a}$$

$$= 2.952 \times 10^{-4} \text{ A.} \tag{S.51}$$

Since there are two unpaired electrons in Ni we can double that number if we like, but since we're just looking for an order-of-magnitude estimate it doesn't really matter either way. Then

$$H = \frac{2.952 \times 10^{-4} \times (10^{-10})^2}{2[(10^{-10})^2 + (3 \times 10^{-10})^2]^{3/2}}$$

$$= 46\,675.7 \frac{\text{A}}{\text{m}} = 586 \text{ Oe.} \tag{S.52}$$

(b) Hund's first rule tells us that the electrons maximize their total spin, S. Therefore they arrange themselves one electron per d orbital with parallel spins before pairing up with opposite spins in the same orbital. For nickel, the resulting configuration looks like this:

Therefore the total spin, $S = 2 \times \frac{1}{2} = 1$.

The five d orbitals have m_l values of $-2, -1, 0, 1$ and 2. Depending on which of the d orbitals contain only one electron, the total M_L can be $-3, -2, -1, 0, 1$ or $2, 3$. Therefore, since $M_L = -L, -L+1, \ldots, 0, \ldots, L-1, L$, the total orbital quantum number, L, must be equal to 3.

Finally, from Hund's third rule, because the shell is more than half full, $J = L + S = 4$.

The allowed values of magnetic moment along the field axis are given by $gM_J\mu_B$, where

$$g = 1 + \frac{J(J+1) + S(S+1) - L(L+1)}{2J(J+1)}$$

$$= 1 + \frac{20 + 2 - 12}{40}$$

$$= 1.25 \tag{S.53}$$

and μ_B is the Bohr magneton. Since $J = 4$, $M_J = -4, -3, -2, -1, 0, 1, 2, 3$ or 4. Therefore the components of the magnetic moment along the field direction can take the values $-5\mu_B, -3.75\mu_B, -2.5\mu_B, -1.25\mu_B, 0, 1.25\mu_B, 2.5\mu_B, 3.75\mu_B$ and $5\mu_B$.

(c) The magnetic dipolar energy,

$$E = -\mu_0 \mathbf{m} \cdot \mathbf{H}. \tag{S.54}$$

Taking $\mathbf{m} = \mu_B$, for a moment aligned as parallel as possible to the field direction, $E = 1.256 \times 10^{-6} \times 5.0 \times -9.274 \times 10^{-24}\,\mathrm{A\,m^2} \times 46\,675.7\,\mathrm{A\,m^{-1}} = -2.72 \times 10^{-24}\,\mathrm{J}$. The energy of a moment aligned as antiparallel as possible to the field is $+2.72 \times 10^{-24}\,\mathrm{J}$. So the difference in magnetic dipole energy between Ni atoms aligned parallel and antiparallel to each other is of the order of $10^{-24}\,\mathrm{J}$.

(d) Below the Curie temperature, T_C, paramagnetic materials exhibit ferromagnetic behavior. Above T_C, the thermal energy outweighs the energy causing ferromagnetic alignment, and the ferromagnetic ordering is destroyed. Therefore the interaction energy which tends to align magnetic moments parallel must be approximately equal to the thermal energy, $k_B T_C = 1.38 \times 10^{-23}\,\mathrm{J\,K^{-1}} \times 631\,\mathrm{K} = 8.7 \times 10^{-21}\,\mathrm{J}$. The magnetic dipole energy is approximately four orders of magnitude smaller than the energy of the *actual* interaction causing the Ni atoms to align ferromagnetically!

(e) The origin of the ferromagnetic coupling in Ni is the quantum mechanical exchange interaction. The exchange interaction is a consequence of the Pauli exclusion principle. If two electrons in an atom have antiparallel spins, then they are allowed to share the same atomic or molecular orbital. As a result they will overlap spatially, thus increasing the electrostatic Coulomb repulsion. In contrast, if they have parallel spins, then they must occupy different orbitals and so will have *less* unfavorable Coulomb repulsion. So the orientation of the spins affects the spatial part of the wavefunction, and this in turn determines the electrostatic Coulomb interaction between the electrons.

(f) In the ferromagnetic transition metals, Fe, Ni and Co, the Fermi energy lies in a region of overlapping 3d and 4s bands, as was shown schematically in Fig. 6.5. As a result of the overlap between the 4s and 3d bands, the valence electrons partially occupy both the 3d and 4s bands. For example, Ni, with 10 valence electrons per atom, has 9.46 electrons in the 3d band and 0.54 electrons in the 4s band. The 4s band is broad, with a low density of states at the Fermi level. Consequently, the energy which would be required to promote a 4s electron into a vacant state so that it could reverse its spin is more than that which would be gained by the resulting decrease in exchange energy. By contrast, the 3d band is narrow and has a much higher density of states at the Fermi level. The large number of electrons near the Fermi level reduces the band energy required to reverse a spin, and the exchange effect dominates. The exchange splitting is negligible for the 4s electrons, but significant for 3d electrons. In Ni, for example, the shift of the bands caused by the exchange interaction is so strong that one 3d sub-band is filled with 5 electrons, and the other contains all 0.54 holes. So the saturation magnetization of Ni is $M = 0.54N\mu_B$, where N is the total number of Ni atoms in the sample.

Chapter 7

7.1　Domains form in ferromagnetic materials in order to minimize the *total* energy. The principle contributions to the magnetic energy of a ferromagnetic material are the exchange energy, which tends to align the magnetic moments parallel to one another, the magnetostatic energy, which is the principal driving force for domain formation, and the magnetocrystalline and magnetostrictive energies, which influence the shape and size of domains.

A magnetized block of ferromagnetic material containing a single domain has a macroscopic magnetization and a magnetic field around it. This causes a magnetostatic energy which can be reduced by dividing the block into domains, as shown schematically in Fig. 7.3.

The tendency of the magnetization to align itself along preferred crystallographic directions in ferromagnetic materials is called the magnetocrystalline anisotropy, and the energy difference between samples magnetized along easy and hard directions is the magnetocrystalline anisotropy energy. To minimize the magnetocrystalline energy, domains form so that their magnetization points along an easy crystallographic direction. In a material with cubic symmetry, both 'vertical' and 'horizontal' directions can be easy axes, therefore the domain arrangement shown in Fig. 7.3(c) has a low magnetocrystalline energy.

When a ferromagnetic material is magnetized it undergoes a change in length known as a magnetostriction. Clearly the horizontal and vertical domains can not elongate (or constrict) at the same time, and instead an elastic strain energy term is added to the total energy. The elastic energy is proportional to the volume of the small perpendicular domains, and can be lowered by reducing the size of these closure domains, which in turn requires smaller primary domains. Of course making smaller domains introduces additional domain walls, with a corresponding increase in energy. The total energy is reduced by a compromise domain arrangement such as that shown in Fig. 7.7.

7.2   Figure 7.11 shows a schematic magnetization curve for a ferromagnetic material, with a sketch of the domain structure at each stage of the magnetization. The magnetic field is applied at an angle which is slightly off the easy axis of magnetization. In the initial demagnetized state, the domains are arranged such that the magnetization averages to zero. When the field is applied, the domain whose magnetization is closest to the field direction starts to grow at the expense of the other domains. The growth occurs by domain wall motion. At first the domain wall motion is reversible; if the field is removed during the reversible stage, the magnetization retraces its path and the demagnetized state is regained. In this region of the magnetization curve the sample does not show hysteresis.

After a while, the moving domain walls encounter imperfections such as defects or dislocations in the crystal. Crystal imperfections have an associated magnetostatic energy. However when a domain boundary intersects the imperfection, this magnetostatic energy can be eliminated, as shown in Fig. 7.12. The intersection of the domain boundary with the imperfection is a local energy minimum. As a result the domain boundary will tend to stay pinned at the imperfection, and energy is required to move it past the imperfection. This energy is provided by the external magnetic field.

The motion of a boundary past an imperfection is shown in Fig. 7.14. When the boundary moves due to a change in the applied field, the domains of closure cling to the imperfection and form spike-like domains. The spike domains persist and stretch as the applied field forces the boundary to move further, until eventually they snap off and the boundary can move freely again. The field required to snap the spike domains off the imperfections corresponds to the coercive force of the material. When the spikes snap from the domain boundary, the discontinuous jump in the boundary causes a sharp change in flux. The change in flux can be observed by winding a coil around the specimen and connecting it to an amplifier and loudspeaker. Even if the applied field is increased very smoothly, crackling noises are heard from the loudspeaker. This phenomenon is known as the Barkhausen effect.

Eventually the applied field is sufficient to eliminate all domain walls from the sample, leaving a single domain, with its magnetization pointing along the easy axis oriented most closely to the external magnetic field. Further increase in magnetization can only occur by rotating the magnetic dipoles from the easy axis of magnetization into the direction of the applied field. In crystals with large magnetocrystalline anisotropy, large fields can be required to reach the saturation magnetization.

As soon as the magnetic field is removed, the dipoles rotate back to their easy axis of magnetization, and the net magnetic moment along the field direction decreases. Since this part of the magnetization process does not involve domain wall motion it is entirely reversible. The demagnetizing field in the sample initiates the growth of reverse magnetic domains which allow the sample to be partially demagnetized. However, the domain walls are unable to fully reverse their motion back to their original positions. This is because the demagnetization process is driven by the demagnetizing field, rather than an applied external field, and the demagnetizing field is not strong enough to overcome the energy barriers when the domain walls encounter crystal imperfections. As a result, the magnetization curve shows hysteresis, and even when the field is removed some magnetization remains in the sample.

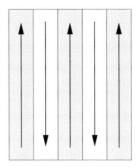

Figure S.4 Domain structure before magnetization of a material with large uniaxial anisotropy.

The coercive field is the additional field, applied in the reverse direction, which is needed to reduce the magnetization to zero.

7.3   Let's assume that the material has a uniaxial anisotropy, so the initial domain structure is as shown in Fig. S.4.

First we'll apply the field along the easy axis (i.e. vertically in the figure). Then the domains which are parallel to the field direction will enlarge by domain wall motion at the expense of those that are antiparallel. Since the material is defect-free it will not exhibit any Barkhausen noise during the magnetization process. The domain wall motion will proceed unimpeded by defects, with the field providing the energy required to rotate each individual magnetic moment out of its initial easy direction, through the hard direction and into the new easy direction. If a material were defect-free and isotropic, it would show no hysteresis. However, for our material with large magnetocrystalline anisotropy, the existence and size of hysteresis depend on the relative magnitudes of the anisotropy and the demagnetizing field at saturation. If the demagnetizing field is large enough to overcome the anisotropy, then domains will start to reform by rotation of the spins from one easy axis, through the hard direction and into the opposite direction. In this case the magnetization will be reversible until the demagnetizing field is no longer large enough to reverse spins on its own, and an external field must be applied to continue the demagnetizing process. If the anisotropy is large enough, however, the demagnetizing field will be too small to spontaneously reform domains, and the material will retain its magnetization until a large enough field is applied in the reverse direction. At this coercive field, there will be a rapid reverse of magnetization, therefore a square hysteresis loop will be formed. The area of the loop will likely be much smaller than that in the same anisotropic material containing defects. If the coercive field is too small, this would be a poor choice for a magnetic data storage medium, since the recorded data bits would not be stable in the presence of small stray fields. The material could be useful in high-frequency applications (such as transformer cores) where the direction of magnetization needs to switch rapidly.

If the field is aligned along the hard axis, then the change of magnetization with applied field is approximately linear, and the retentivity and coercivity are close to zero. Such a material could be used whenever a linear *M*–*H* curve is required.

A polycrystalline sample will have a behavior between these two extremes.

7.4   Materials with high defect content show Barkhausen noise during the magnetization process. Large fields must be applied to move the domain wall past the defects and achieve saturation, therefore they are hard magnetic materials. After saturation, when the field is removed, the defects resist the reformation of domain walls, therefore these materials have large area hysteresis loops with large remanence and high coercive field. Magnetically hard materials with many defects are used as permanent magnets.

7.5   At the origin, when B and H are both equal to zero, domains are aligned in opposite directions such that the total magnetization is zero. As the field is increased, domains aligned closest to the field direction grow by domain wall motion at the expense of the other domains until eventually a single domain is formed. The saturation induction is reached when the magnetization direction of this domain rotates into the direction of the applied field. As the external field is reduced to zero, the demagnetizing field causes domains of reverse magnetization to nucleate and the net magnetization is reduced. As the field is increased in the opposite direction the domains of reverse magnetization grow. At H_c, the induction is zero, but there is still a small positive magnetization, since $B = H + 4\pi M = 0$, so $M = -H_c/4\pi$. At this point, prior to saturation in the reverse direction, the magnetizing field is reversed once again, and the minor hysteresis loop is traced out. When the magnetizing field is reduced from its negative value to zero, the resulting induction is less than the remanent induction because the starting point was not the saturation induction. The field is then re-applied in the negative direction and increased to the value of the coercive field, at the starting point of the minor hysteresis loop. Just as in the initial magnetization process, domains which are aligned closest to the field direction are expanded and rotated at each stage.

7.6   The demagnetizing path falls progressively shorter of saturation at each field reversal. The field is not taken far enough to reach saturation and so some oppositely oriented domains remain; as a result, fewer domains are reoriented along the field direction each time. There is increasingly less alignment of the domains and therefore a reduced magnetization. An alternative way of converting a ferromagnetic material into an unmagnetized state is by heating it above the Curie temperature.

7.7(a)   The exchange energy cost, σ_{ex}, anisotropy energy cost, σ_A, and total energy cost of domain wall formation are plotted in Fig. S.5.

(b)   The minimum energy occurs when $d(\sigma_{ex} + \sigma_A)/dN = 0$. That is $(-k_B T_C/2) \times (\pi/a)^2(1/N^2) + Ka = 0$. Solving for N gives

$$N = \frac{\pi}{a}\sqrt{\left(\frac{k_B T_C}{2Ka}\right)}.$$

(Note that this also corresponds to the N value for which $\sigma_{ex} = \sigma_A$). Therefore the number of layers,

$$N + 1 = \frac{\pi}{a}\sqrt{\left(\frac{k_B T_C}{2Ka}\right)} + 1.$$

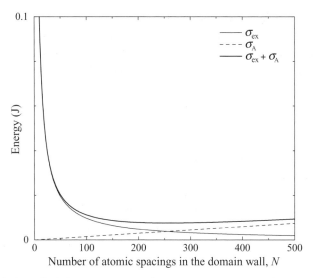

Figure S.5 Variation of exchange energy, anisotropy energy and their sum as a function of domain wall thickness.

(c) Substituting the values of K, T_C and a for iron into this expression gives $N = 229$. Therefore the wall thickness, $Na = 68.7$ nm, and the wall energy is 0.007 J/m^2.

Chapter 8

8.1 The susceptibility of an antiferromagnetic material in which the field is applied parallel to the magnetization direction is given by

$$\chi_\parallel = \frac{2Nm^2 B'(J, \alpha)}{2k_B T + Nm^2 \gamma B'(J, \alpha)}. \tag{S.55}$$

Here $B'(J, \alpha)$ is the derivative with respect to α of the Brillouin function,

$$B_J(\alpha) = \frac{2J+1}{2J} \coth\left(\frac{2J+1}{2J}\alpha\right) - \frac{1}{2J} \coth\left(\frac{\alpha}{2J}\right), \tag{S.56}$$

and $\alpha = Jg\mu_B H/k_B T$.

At high temperature, α is very small, therefore we can expand the Brillouin function in a Taylor series about the origin to obtain

$$B_J(\alpha) = \frac{J+1}{3J}\alpha - \frac{[(J+1)^2 + J^2](J+1)}{90J^3}\alpha^3 + \cdots. \tag{S.57}$$

So, at small α, $B'(J, \alpha) = (J+1)/3J$, which is a constant for a given J. Then

$$\chi_\parallel = \frac{2Nm^2(J+1)/3J}{2k_B T + Nm^2\gamma(J+1)/3J} \tag{S.58}$$

$$= \frac{C}{T+\theta} \tag{S.59}$$

– the Curie–Weiss law!

At low temperature, α is large and we cannot use the expansion of the Brillouin function given above. Instead, using the fact that $d(\coth\alpha)/d\alpha = -\operatorname{cosech}^2\alpha$, we differentiate the Brillouin function explicitly to obtain

$$B'(J,\alpha) = -\left(\frac{2J+1}{2J}\right)^2 \operatorname{cosech}^2\left(\frac{2J+1}{2J}\alpha\right) + \left(\frac{1}{2J}\right)^2 \operatorname{cosech}^2\left(\frac{\alpha}{2J}\right). \quad (S.60)$$

As $\alpha \to \infty$, $\operatorname{cosech}(\alpha) \to 0$, and $B'(J,\alpha) \to 0$. Therefore $\chi_\parallel$ also tends to zero at low temperature.

8.2   Since the A–B interaction is much stronger than the A–A and B–B interactions, we can use the results which we derived in the notes using the Langevin–Weiss theory. We know that the expression for the susceptibility at and above the Néel temperature is

$$\chi = \frac{C}{T+\theta} = \frac{C}{T+T_N}. \quad (S.61)$$

In this case, we're told that $\chi(T_N) = \chi_0$, so we can solve for the constant, C, to obtain $C = 2T_N\chi_0$. Then, at $T = 2T_N$, $\chi = C/(2T_N + T_N) = 2T_N\chi_0/3T_N = \frac{2}{3}\chi_0$. Below T_N, for the field applied perpendicular to the magnetization, χ is a constant equal to its value at T_N. Therefore at both $T = 0$ and $T = T_N/2$, $\chi = \chi_0$.

Chapter 9

9.1 Review question 1

(a) Ferrimagnets behave similarly to ferromagnets, in that they exhibit a spontaneous magnetization below some critical temperature, T_C, even in the absence of an applied field. Their permeabilities and susceptibilities are large and positive, and they concentrate magnetic flux within themselves. Both tend to form domains in the spontaneously magnetized phase. However, the detailed shape of the ferrimagnetic magnetization curve is distinctly different from that of the ferromagnetic curve. The reason for this is that the local arrangement of magnetic moments is quite different. In ferromagnets, adjacent moments align parallel, whereas ferrimagnets consist of two interpenetrating sublattices with opposite alignment of magnetic moments, but the magnetizations of the two sublattices are different giving a net magnetic moment. Most ferrimagnets are ionic solids, whereas most ferromagnets are metals, so the electrical properties of ferrimagnets are quite different from those of ferromagnets. This results in a wide range of important applications for ferrimagnets, in situations requiring magnetic insulators.

(b) The measured spontaneous magnetization of magnetite is plotted as a function of temperature in Fig. S.6. The results agree well with the classical ($J = \infty$) magnetization curve predicted for *ferromagnets* within the Langevin–Weiss theory! In this case the agreement is fortuitous, but historically it led Weiss and his co-workers to believe that magnetite was a ferromagnetic material, and gave them great confidence in the localized moment theory.

(c) The saturation magnetization is defined to be the magnetic moment per unit volume. Therefore we'll calculate the magnetic moment of a unit cell of Fe_3O_4, and divide by the unit cell volume, which is $(0.83 \times 10^{-9})^3$ m^3. In ferrites, the magnetic moments of the

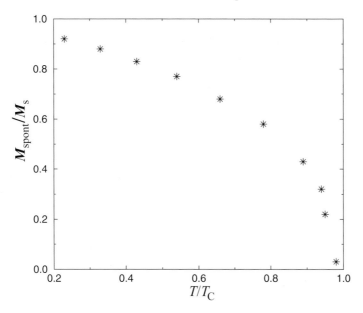

Figure S.6 Spontaneous magnetization as a function of temperature for magnetite.

Fe^{3+} ions cancel out, so the net magnetic moment arises from the Fe^{2+} ions only. Each Fe^{2+} ion has a magnetic moment of $4\mu_B$, since there are six 3d electrons, two of which occupy the same orbital with their spins opposed, leaving four uncompensated parallel spins. There are eight Fe^{2+} ions per unit cell, so the total magnetic moment per unit cell is $32\mu_B$. Then the saturation magnetization is

$$M_s = \frac{32 \times 9.27 \times 10^{-24}}{(0.839 \times 10^{-9})^3} \frac{\text{A m}^2}{\text{m}^3}$$

$$= 5.0 \times 10^5 \text{A/m}. \tag{S.62}$$

(d) In part (c) we found that the saturation magnetization of Fe_3O_4 is 5.0×10^5 A/m. In order to increase the magnitude of M_s, we need to replace some fraction of the Fe^{2+} ions with divalent metal ions that have a larger magnetic moment. Our only option from the 3d transition series is Mn^{2+}, which has a moment of 5 Bohr magnetons per atom (as opposed to 4 in Fe^{2+}.) If we assume that the unit cell size does not change when we substitute Mn^{2+} ions for Fe^{2+} ions, then we can calculate the number of Bohr magnetons per unit cell which this saturation magnetization corresponds to:

$$\text{number of Bohr magnetons per unit cell} = \frac{M_s \times \text{unit cell volume}}{\mu_B}$$

$$= \frac{(5.25 \times 10^5 \text{ A/m})(0.839 \times 10^{-9})^3 \text{ m}^3}{9.27 \times 10^{-24} \text{ A m}^2}$$

$$= 33.45 \text{ Bohr magnetons per unit cell.}$$

Let the fraction of Mn^{2+} ions be x, and the fraction of Fe^{2+} ions be $(1-x)$. Then, since there are eight divalent ions per unit cell,

$$8[5x + 4(1 - x)] = 33.45, \tag{S.63}$$

so $x = 0.181$. Therefore if we replace 18.1% of the Fe^{2+} ions in Fe_3O_4 by Mn^{2+} ions, the saturation magnetization will be increased to 5.25×10^5 A/m.

The saturation flux density, $B_s = \mu_0 M_s = 4\pi \times 10^{-7}$ H/m $\times 5.25 \times 10^5$ A/m $= 0.66$ T. To convert to cgs units we use the fact that 1 gauss $= 10^{-4}$ teslas so the saturation flux density is 6.6×10^3 gauss.

(e) Ferrimagnets are related to antiferromagnets, in that the exchange coupling between adjacent magnetic ions leads to antiparallel alignment below some critical temperature. Both are paramagnetic above the critical temperature, although the details of their susceptibility curves differ. Below the critical temperature antiferromagnets have no net magnetization. In contrast, ferrimagnets have a net magnetization because the magnetization of one sublattice is greater than that of the oppositely oriented sublattice.

(f) The bonding in ferrimagnetic materials is largely ionic, with transition metal cations having oxygen anions as their nearest neighbors. The d electrons on the transition metal ions obey Hund's rules, and occupy the five d orbitals singly with spins parallel before pairing up.

We make the assumption that it is energetically favorable for the valence electrons on the cations to undergo some degree of covalent bonding with those on the O^{2-} ions. Since the O^{2-} ion has a filled shell of electrons, this can only take place by donation of electrons from the O^{2-} ion into the vacant orbitals of the cation. As an example, let's assume that our left-most cation is an up-spin Mn^{2+} ion, as shown in Fig. 8.14. Then, since all the Mn orbitals contain an up-spin electron, covalent bonding can only occur if the neighboring oxygen donates its down-spin electron. This leaves an up-spin electron in the oxygen p orbital, which it is able to donate to the next cation in the chain. (In Fig. 8.14 this is another Mn^{2+} ion.) By the same argument, bonding can only occur if the electrons on the next cation are down-spin. We see that this oxygen-mediated interaction leads to an overall antiferromagnetic alignment between the cations, without requiring the quantum mechanical exchange integral to be negative!

Since the superexchange interaction relies on overlap between the O 2p orbitals and the neighboring transition metal cations, which is largest in linear cation–oxygen–cation chains, the strength of the superexchange interaction is reduced if the cation–oxygen–cation bond angle is changed from 180°.

9.2 Review question 2

(a) The electronic configuration of an Fe atom is

$$(1s)^2(2s)^2(2p)^6(3s)^2(3p)^6(4s)^2(3d)^6.$$

The iron ion in Fe_2O_3 is a trivalent cation. Therefore, because the transition elements give up their 4s electrons before their 3d electrons on ionization, the electronic configuration of

an Fe^{3+} *ion* is

$$(1s)^2(2s)^2(2p)^6(3s)^2(3p)^6(3d)^5.$$

The electronic configuration of a Ni atom is

$$(1s)^2(2s)^2(2p)^6(3s)^2(3p)^6(4s)^2(3d)^8.$$

The nickel ion in NiO is a divalent cation with electronic configuration

$$(1s)^2(2s)^2(2p)^6(3s)^2(3p)^6(3d)^8.$$

(b) The cations in tetrahedral sites are bonded via O^{2-} ions to cations in octahedral sites. Although the inter-ion interactions in ferrites are largely ionic, the energy of the system can be lowered by some degree of covalent bonding. When covalent bonding occurs, the up-spin (say) cation in the tetrahedral site overlaps with the down-spin 2p electron in the oxygen orbital pointing towards the cation. This leaves the up-spin 2p orbital to bond with the neighboring cation in the octahedral site. A covalent bond with the second cation can only be formed if this cation is down-spin. This mechanism driving antiferromagnetic ordering in predominantly ionic materials is called *superexchange*. Since the iron ions are equally distributed between the octahedral and tetrahedral sites, there are equal numbers of up- and down-spin iron ions, and the net magnetic moment from the iron ions is zero.

(c) Remember that the saturation magnetization is the magnetic moment per unit volume. Therefore we need to work out the magnetic moment and the volume of one unit cell, and take the ratio.

The volume of the unit cell is $(8.34 \times 10^{-10})^3$ m^3 since the unit cell is cubic.

Hund's first rule tells us that the electrons maximize their total spin, S. Therefore they arrange themselves one electron per d orbital with parallel spins before pairing up with opposite spins in the same orbital. For Ni^{2+}, the resulting configuration looks like this:

Since the 3d transition elements have strong quenching of the orbital angular momentum, we only need to consider the spin contribution to the magnetic moment, which we can see from the figure is $2\mu_B$ per atom along the direction of applied field. Finally, since there are eight Ni^{2+} ions per unit cell, the magnetic moment per unit cell is $16\mu_B$.

So

$$M_s = \frac{16\mu_B}{(8.34 \times 10^{-10})^3 \text{ m}^3}$$

$$= \frac{16 \times 9.27 \times 10^{-24} \text{ A m}^{-2}}{(8.34 \times 10^{-10})^3 \text{ m}^3}$$

$$= 2.56 \times 10^5 \text{ A/m}.$$

(d) Now that we are talking about Ni *metal*, we have to worry about overlapping bands. If the number of free electrons per atom is 0.54, then the number of s electrons per Ni atom must also equal 0.54. But we know that the number of valence electrons in a Ni atom is 10. Therefore the number of d electrons per Ni atom must just be equal to the difference – that is 9.46.

(e) Since it takes 5 electrons per Ni atom to completely fill the up-spin band, the remaining 4.46 electrons per atom go into the down-spin band. Therefore the net magnetic moment per Ni atom, which is just the number of up-spin electrons minus the number of down-spin electrons times μ_B, is $0.54\mu_B$. The density of states of ferromagnetic nickel is shown below.

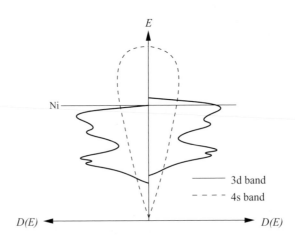

(f) There are four atoms per unit cell in the fcc structure (one at the center of the unit cell, eight corner atoms which are shared between eight unit cells, and six face-centered atoms each shared between two unit cells). Therefore the magnetic moment per unit cell is $8 \times 0.54\mu_B$. Since the volume of the unit cell is $(3.52 \times 10^{-10})^3$ m^3, the saturation magnetization of elemental nickel,

$$M_s = \frac{2.16\mu_B}{(3.52 \times 10^{-10})^3 \text{ m}^3} = 4.59 \times 10^5 \text{ A/m}.$$

(g) Note that the saturation magnetization of Ni is greater than that of nickel ferrite, even though the number of Bohr magnetons per atom is smaller. This is because *all* of the atoms in elemental Ni contribute to the magnetization, whereas many of the atoms in nickel ferrite are either non-magnetic, or have magnetic moments which cancel with those of neighboring ions. Different applications for Ni and Ni-ferrite result from their different electrical properties. Ni-ferrite is an insulator and therefore could be useful for high-frequency applications, such as transformer cores. Also, because of its anisotropy, it could be a good storage medium in magnetic memory applications. Nickel, with its higher saturation magnetization, would be a better material for permanent magnets and electromagnets.

Chapter 10

10.1 Remember that, in all cases, the stable domain structure minimizes the *total* energy of the system.

If a material had no magnetocrystalline anisotropy, then there would be no preferred direction of alignment of the magnetic moments. Therefore it would be possible to eliminate the magnetostatic energy without domain formation using the spin configuration shown in Fig. S.7(a). This would be a favorable arrangement since adjacent spins are still parallel, optimizing the exchange energy, and magnetostrictive energy is not introduced.

A large uniaxial anisotropy causes the magnetic moments to be aligned along a single crystallographic direction. So, 90° domain walls and perpendicular domains of closure are unlikely. A probable domain structure is shown in Fig. S.7(b).

The magnetostrictive energy can be minimized by increasing the volume of the main domains which have their magnetizations parallel to a certain easy direction, as shown in Fig. S.7(c). This arrangement concentrates the elastic energy into the small flux-closure domains which are forced to strain to fit the deformation of the main domains.

If the sample is smaller than the domain wall thickness, then domain formation becomes energetically unfavorable. In this case the particle will consist of a single domain, as shown in Fig. S.7(d).

10.2 In a single-domain particle, the magnetization lies along an easy direction which is determined by magnetocrystalline and shape anisotropies. If a field is applied in the direction opposite to the magnetization (but still in the easy direction), then the particle cannot respond by domain wall motion, and the magnetization must *rotate* through the hard direction in order to reverse its direction. Anisotropy forces tend to hold the magnetization in the easy direction therefore the coercivity is large. A square hysteresis loop is produced. If the field is applied along a hard direction, the magnetization rotates into the field direction when a large enough field is applied, but rotates fully back to the easy direction when the field is removed. Therefore there is no hysteresis. Small-particle magnets are used for recording media where a high coercivity is required. Usually needle-shaped particles are used in order to maximize the shape anisotropy and increase the coercive force. The particles must be aligned with their easy axes parallel to the direction in which the write field will be applied.

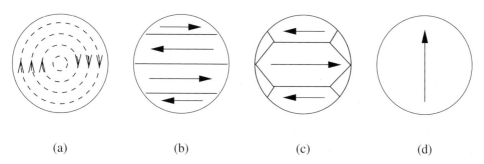

(a) (b) (c) (d)

Figure S.7 Domain arrangements for four different hypothetical materials.

Chapter 11

11.1 Review question

(a) This is the same problem as Exercise 1.3a. Using the Biot–Savart law we obtain the following expression for the field generated on the axis of a circular coil.

$$H_{\text{axial}} = \frac{Ia^2}{2(a^2 + x^2)^{3/2}}.$$

Using the given value for angular momentum we can evaluate the current, and obtain a value for the field

$$H = 46\,675.7\,\text{A m}^{-1} = 586\,\text{Oe}.$$

(b) This is the same problem as Exercise 1.3b. The magnetic dipole moment, m, is given by

$$m = IA$$
$$= 9.274 \times 10^{-24}\,\text{A m}^2 \text{ or J T}^{-1}$$

i.e. one Bohr magneton.

(c) The field lines around the dipole, oriented with its north pole upwards, are shown below.

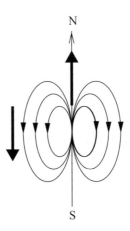

If a second dipole is directly above the first, the field from the first dipole will tend to align it vertically, with its north pole pointing upwards. If a second dipole is positioned horizontally from the original dipole, it will again be aligned vertically, but with its north pole pointing downwards.

(d) The magnetic ordering is shown below.

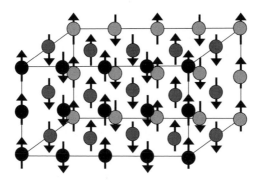

Here the black atoms are in the front-most plane, the dark-gray atoms are in the middle plane, and the light-gray atoms are in the back plane.

(e) If the second dipole is aligned parallel to the first, its energy will be *lowered* by an amount

$$E = -\mu_0 \boldsymbol{m} \cdot \boldsymbol{H}$$
$$= 1.256 \times 10^{-6}\,\text{Wb A}^{-1}\,\text{m}^{-1} \times -9.274 \times 10^{-24}\,\text{A m}^2 \times 46\,675.7\,\text{A m}^{-1}$$
$$= 5.44 \times 10^{-25}\,\text{J}.$$

If it is aligned antiparallel its energy will be raised by the same amount. This magnetic energy corresponds to a temperature, $T = E/k_B = 0.0394$ K. Note that this number is very small, so it is unlikely that the parallel alignment of magnetic dipole moments in ferromagnetic materials results from a magnetic dipolar interaction.

(f) The electronic structures of the Mn ions are:

$$\text{Mn}^{3+}\ [\text{Ar}]\ (3d)^4$$
$$\text{Mn}^{4+}\ [\text{Ar}]\ (3d)^3$$

Assuming spin-only magnetic moments, then the Mn^{3+} ion has a maximum magnetic moment along the field direction of $4\mu_B$, and the Mn^{4+} ion has a corresponding magnetic moment of $3\mu_B$.

(g) There is an excellent discussion of the relationship between chemical bonding and magnetic ordering in manganites in the landmark 1955 paper by Goodenough.[54] Here we follow Goodenough's argument.

(i) In LaMnO_3, all of the manganese ions are Mn^{3+}, with four 3d electrons. Following Hund's rule, the four 3d electrons each occupy a different 3d orbital so that they can align parallel to each other. This leaves one vacant 3d orbital. The oxygen-mediated coupling between neighboring manganese ions can be either ferromagnetic or antiferromagnetic, depending on whether empty or filled manganese d orbitals point towards the oxygen. Figure S.8(a) illustrates the antiferromagnetic superexchange which results when both

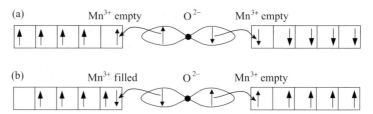

Figure S.8 Superexchange between empty and filled Mn^{3+} orbitals, resulting in ferromagnetic coupling between the Mn ions.

Mn^{3+} ions have an empty d orbital pointing towards the oxygen anion. In this case, the left-most Mn^{3+} ion is up-spin, and so the up-spin oxygen 2p electron donates into the empty 3d orbital in order to optimize its Hund's rule coupling to the manganese ion. This leaves the down-spin oxygen p electron available to donate into the right-hand manganese. The Hund's rule coupling is optimized if this second Mn^{3+} ion is down-spin, that is antiferromagnetically aligned with the first manganese ion.

The opposite situation, where the oxygen anion links one empty and one filled Mn 3d orbital, is shown in Fig. S.8(b). As before, the up-spin oxygen electron donates into the empty d orbital of the up-spin Mn^{3+} ion. The down-spin oxygen electron can only interact with its neighboring filled Mn 3d orbital by covalent bond formation, which can only occur if the Mn 3d electron has the *opposite* (i.e. up-) spin. Therefore the second manganese ion must have the same spin orientation as the first, resulting in ferromagnetic coupling.

Valence bond theory tells us that the single empty d orbital in a Mn^{3+} ion will hybridize with the Mn 4s and 4p orbitals to form four square planar dsp^2 empty orbitals. Since the Mn ions in $LaMnO_3$ are octahedrally coordinated, this means that only $\frac{2}{3}$ of the bonds to oxygen can be empty and hence each Mn ion is bonded ferromagnetically to four of its neighbors, and antiferromagnetically to two. Since the ferromagnetic bonds are longer than the antiferromagnetic bonds, orbital ordering occurs to minimize the elastic strain in the lattice. The result of this is the so-called A-type antiferromagnetic ordering, that is ferromagnetic planes of Mn ions with adjacent planes aligned antiferromagnetically to each other.

(ii) In $CaMnO_3$, all of the ions are Mn^{4+}, with two empty d orbitals per Mn ion. These two empty d orbitals hybridize with the Mn 4s and 4p orbitals to form *six* octahedral d^2sp^3 empty orbitals. Therefore *all* bonds to oxygen can be made by empty Mn d orbitals, resulting in the so-called G-type antiferromagnetism, in which all bonds are antiferromagnetic.

The magnetic dipole energy calculated above is the equivalent of the order of *hundredths* of kelvin, five orders of magnitude smaller than the actual ordering temperature of $CaMnO_3$. This implies that the superexchange interaction mechanism just described, which is responsible for the antiferromagnetic ordering in $CaMnO_3$, is correspondingly five orders of magnitude stronger than the dipole–dipole interaction between neighboring Mn ions.

(h) Adjacent Mn^{3+} and Mn^{4+} ions are coupled by the so-called *double exchange* mechanism which results in ferromagnetic coupling.[57] The total energy of the Mn^{3+}–Mn^{4+} pair can be lowered if the extra 3d electron on the Mn^{3+} ion is allowed to resonate or tunnel between the two ions in the pair. (This is analogous to the lowering in energy of the ground state of an ammonia molecule by inversion tunneling). Electron tunneling can only occur if the magnetic moments on both Mn ions are aligned parallel to each other, so that the up-spin (say) electron on the Mn^{3+} ion is able to transfer to the Mn^{4+} ion and be parallel to the 3d electrons on its new host. This mechanism is called double exchange because the electron in fact transfers from the Mn^{3+} ion to the intermediate oxygen, simultaneously with the transfer of an electron from the O^{2-} ion to the Mn^{4+} ion.

References

[1] W.F. Brown Jr. Tutorial paper on dimensions and units. *IEEE Trans. Mag.*, **20** 112, 1984.

[2] P. Hammond. *Electromagnetism for engineers*. Pergamon Press, 1978.

[3] C.W. Trowbridge. Electromagnetic computing: the way ahead? *IEEE Trans. Mag.*, **24** 13, 1988.

[4] R.P. Feynman, R.B. Leighton, and M. Sands. *The Feynman lectures on physics*. Addison-Wesley, Reading, MA, 1965.

[5] P. Zeeman. Influence of magnetism on the nature of the light emitted by a substance. *Phil. Mag.*, **5** 226, 1897.

[6] P.W. Atkins. *Molecular quantum mechanics*. Oxford University Press, 1999.

[7] H.N. Russell and F.A. Saunders. New regularities in the spectra of the alkaline earths. *Astrophys. J.*, **61** 38, 1925.

[8] F. Hund. *Linienspektren und periodische system der elemente*. Springer, Berlin, 1927.

[9] F. Paschen and E. Back. Normale und anomale zeemaneffekte. *Annln. Phys.*, **40** 960, 1913.

[10] P. Langevin. Magnétisme et théorie des électrons. *Annales de Chemie et de Physique*, **5** 70, 1905.

[11] W. Pauli. *Z. Phys.*, **2** 201, 1920.

[12] A. Firouzi, D.J. Schaefer, S.H. Tolbert, G.D. Stucky, and B.F. Chmelka. Magnetic-field-induced orientational ordering of alkaline lyotropic silicate–surfactant liquid crystals. *J. Am. Chem. Soc.*, **119** 9466, 1997.

[13] S.H. Tolbert, A. Firouzi, G.D. Stucky, and B.F. Chmelka. Magnetic field alignment of ordered silicate–surfactant composites and mesoporous silica. *Science*, **278** 264, 1997.

[14] W. Meissner and R. Ochsenfeld. *Naturwiss.*, **21** 787, 1933.

[15] B.D. Josephson. Possible new effects in superconductive tunneling. *Phys. Lett.*, **1** 251, 1962.

[16] C. Kittel. *Introduction to solid state physics*. John Wiley & Sons, 1996.

[17] M.L. Cohen. The pseudopotential panacea. *Physics Today* (July) 40, 1979.

[18] P. Weiss. L'hypothèse du champ moléculaire et la propriété ferromagnétique. *J. Phys.*, **6** 661, 1907.

[19] F. Tyler. The magnetization–temperature curves of iron, cobalt and nickel. *Phil. Mag.*, **11** 596, 1931.

[20] W. Heisenberg. On the theory of ferromagnetism. *Z. Phys.*, **49** 619, 1928.

[21] J.C. Slater. Electronic structure of alloys. *J. Appl. Phys.*, **8** 385, 1937.

[22] L. Pauling. The nature of the interatomic forces in metals. *Phys. Rev.*, **54** 899, 1938.

[23] D.J. Singh, W.E. Pickett, and H. Krakauer. Gradient-corrected density functionals: full-potential calculations for iron. *Phys. Rev. B*, **43** 11628, 1991.

[24] H.J.F. Jansen. Electronic structure calculations for magnetically ordered systems. *Physics Today* (April) 50, 1995.

[25] F. Bitter. On inhomogeneities in the magnetization of ferromagnetic materials. *Phys. Rev.*, **38** 1903, 1931.

[26] H.J. Williams, F.G. Foster, and E.A. Wood. Observation of magnetic domains by the Kerr effect. *Phys. Rev.*, **3** 119, 1951.

[27] C.A. Fowler and E.M. Fryer. Magnetic domains by the longitudinal Kerr effect. *Phys. Rev.*, **94** 52, 1954.

[28] H.J. Williams, R.M. Bozorth, and W. Shockley. Magnetic domain patterns on single crystals of silicon iron. *Phys. Rev.*, **75** 155, 1949.

[29] H. Barkhausen. Two phenomena uncovered with the help of new amplifiers. *Z. Phys.*, **20** 401, 1919.

[30] J.F. Dillon Jr. Observation of domains in the ferrimagnetic garnets by transmitted light. *J. Appl. Phys.*, **29** 1286, 1958.

[31] G. Shull and J.S. Smart. Detection of antiferromagnetism by neutron diffraction. *Phys. Rev.*, **76** 1256, 1949.

[32] G.E. Bacon. *Neutron diffraction*. Oxford: Clarendon Press, 1975.

[33] L. Néel. Propriétés magnétique des ferrites: ferrimagnétisme et antiferromagnétisme. *Annales de Physique*, **3** 137, 1948.

[34] B.D. Cullity. *Introduction to magnetic materials*. Addison-Wesley, 1972.

[35] J. Rath and J. Callaway. Energy bands in paramagnetic chromium. *Phys. Rev. B*, **8** 5398, 1973.

[36] M.A. Ruderman and C. Kittel. Indirect exchange coupling of nuclear magnetic moments by conduction electrons. *Phys. Rev.*, **96** 99, 1954.

[37] T. Kasuya. Electrical resistance of ferromagnetic metals. *Prog. Theor. Phys.*, **16** 58, 1956.

[38] K. Yosida. Magnetic properties of Cu–Mn alloys. *Phys. Rev.*, **106** 893, 1957.

[39] W.H. Meikeljohn and C.P. Bean. New magnetic anisotropy. *Phys. Rev.*, **105** 904, 1957.

[40] J. Nogués and I.K. Schuller. Exchange bias. *J. Mag. Mag. Mat.*, **192** 203, 1999.

[41] A. Serres. Récherches sur les moments atomiques. *Annales de Physique*, **17** 5, 1932.

[42] O. Kahn. The magnetic turnabout. *Nature*, **399** 21, 1999.

[43] S. Ohkoshi, Y. Abe, A. Fujishima, and K. Hashimoto. Design and preparation of a novel magnet exhibiting two compensation temperatures based on molecular field theory. *Phys. Rev. Lett.*, **82** 1285, 1999.

[44] H. van Leuken and R.A. de Groot. Half-metallic antiferromagnets. *Phys. Rev. Lett.*, **74** 1171, 1995.

[45] W.E. Pickett. Spin-density-functional-based search for half-metallic antiferromagnets. *Phys. Rev. B*, **57** 10613, 1998.

[46] C. Kittel, J.K. Galt, and W.E. Campbell. Crucial experiment demonstrating single domain property of fine ferromagnetic powders. *Phys Rev.*, **77** 725, 1950.

[47] C.P. Bean and I.S. Jacobs. Magnetic granulometry and super-paramagnetism. *J. Appl. Phys.*, **27** 1448, 1956.

[48] W.P. Jayasekara, S. Wang, and M.H. Kryder. 4 Gbit/in^2 inductive write heads using high moment FeAlN poles. *J. Appl. Phys.*, **79** 5880, 1996.

[49] J. Kondo. Anomalous Hall effect and magnetoresistance of ferromagnetic metals. *Prog. Theor. Phys.*, **27** 772, 1962.

[50] M.N. Baibich, J.M. Broto, A. Fert, F. Nguyen Van Dau, F. Petroff, P. Etienne, G. Creuzet, A. Friederich, and J. Chazelas. Giant magnetoresistance of (001)Fe/(001)Cr magnetic superlattices. *Phys. Rev. Lett.*, **61** 2472, 1988.

[51] G.A. Prinz. Magnetoelectronics. *Science*, **282** 1660, 1998.

[52] S. Jin, T.H. Tiefel, M. McCormack, R.A. Fastnacht, R. Ramesh, and L.H. Chen. Thousandfold change in resistivity in magnetoresistive La–Ca–Mn–O films. *Science*, **264** 413, 1994.

[53] G. Xiao, A. Gupta, X.W. Li, G.Q. Gong, and J.Z. Sun. Sub-200 Oe giant magnetoresistance in manganite tunnel junctions. *Proc. Symp. Sci. Tech. Magnetic Oxides, Boston, MA, 1–4 Dec. 1997*, Materials Research Society, 1998, p. 221.

[54] J.B. Goodenough. Theory of the role of covalence in the perovskite-type manganites [LaM(II)]MnO$_3$. *Phys. Rev.*, **100** 564, 1955.

[55] E.O. Wollan and W.C. Koehler. Neutron diffraction study of the magnetic properties of the series of perovskite-type compounds [La$_{1-x}$Ca$_x$]MnO$_3$. *Phys. Rev.*, **100** 545, 1955.

[56] P. Schiffer, A.P. Ramirez, W. Bao, and S.-W. Cheong. Low temperature magnetoresistance and the magnetic phase diagram of La$_{1-x}$Ca$_x$MnO$_3$. *Phys. Rev. Lett.*, **75** 3336, 1995.

[57] C. Zener. Interaction between the d shells in transition metals II. Ferromagnetic compounds of manganese with perovskite structure. *Phys. Rev.*, **82** 403, 1951.

[58] D.A. Thompson and J.S. Best. The future of magnetic data storage technology. *IBM J. Res. Develop.*, **44** 311, 2000.

[59] C.B. Murray, S. Shouheng, H. Doyle, and T. Betley. Monodisperse 3d transition-metal (Co, Ni, Fe) nanoparticles and their assembly into nanoparticle superlattices. *MRS Bulletin*, **26** 985, 2001.

[60] R.W. Cross, J.O. Oti, S.E. Russek, T. Silva, and Y.K. Kim. Magnetoresistance of thin-film NiFe devices exhibiting single-domain behavior. *IEEE Trans. Mag.*, **31** 3358, 1995.

[61] T. Suzuki. Magneto-optic recording materials. *MRS Bulletin*, **21** (9) 42, 1996.

[62] R.J. Gambino and T. Suzuki. *Magneto-optical recording materials*. John Wiley, 1999.

[63] N. Samarth, P. Klosowski, H. Luo, T.M. Giebultowicz, J.K. Furdyna, J.J. Rhyne, B.E. Larson, and N. Otsuka. Antiferromagnetism in ZnSe/MnSe strained-layer superlattices. *Phys. Rev. B*, **44** 4701, 1991.

[64] J.K. Furdyna. Diluted magnetic semiconductors. *J. Appl. Phys.*, **64** R29, 1988.

[65] J.K. Furdyna. Diluted magnetic semiconductors – an interface of semiconductor physics and magnetism. *J. Appl. Phys.*, **53** 7637, 1982.

[66] S.A. Crooker, D.A. Tulchinsky, J. Levy, D.D. Awschalom, R. Garcia, and N. Samarth. Enhanced spin interactions in digital magnetic heterostructures. *Phys. Rev. Lett.*, **75** 505, 1995.

[67] S.A. Crooker, D.D. Awschalom, J.J. Bamuberg, F. Flack, and N. Samarth. Optical spin resonance and transverse spin relaxation in magnetic semiconductor quantum wells. *Phys. Rev. B*, **56** 7574, 1997.

[68] M.A. Nielsen and I.L. Chuang. *Quantum computation and quantum information*. Cambridge University Press, 2001.

[69] I.P. Smorchkova, N. Samarth, J.M. Kikkawa, and D.D. Awschalom. Spin transport and localization in a magnetic two-dimensional electron gas. *Phys. Rev. Lett.*, **78** 3571, 1997.

[70] I. Smorchkova and N. Samarth. Fabrication of n-doped magnetic semiconductor heterostructures. *Appl. Phys. Lett.*, **69** 1640, 1996.

[71] H. Ohno. Making nonmagnetic semiconductors ferromagnetic. *Science*, **281** 951, 1998.

[72] T. Sasaki, S. Sonoda, Y. Yamamoto, K. Suga, S. Shimizu, and H. Hori. Molecular beam epitaxy of wurtzite (Ga,Mn)N films on sapphire(0001) showing the ferromagnetic behaviour at room temperature. *J. Appl. Phys.*, **91** 7911, 2002.

[73] S. Sanvito, G. Theurich, and N.A. Hill. Density functional calculations for III–V diluted ferromagnetic semiconductors: a review. *J. Supercon.*, **15** 85, 2002.

[74] T. Dietl, H. Ohno, F. Matsukura, J. Cibèrt, and D. Ferrand. Zener model description of ferromagnetism in zinc-blende magnetic semiconductors. *Science*, **287** 1019, 2001.

[75] T. Jungwirth, W.A. Atkinson, B.H. Lee, and A.H. MacDonald. Interlayer coupling in ferromagnetic semiconductor superlattices. *Phys. Rev. B*, **59** 9818, 1999.

[76] T. Hayashi, Y. Hashimoto, S. Katsumoto, and Y. Iye. Effect of low-temperature annealing on transport and magnetism of diluted magnetic semiconductor (Ga,Mn)As. *Appl. Phys. Lett.*, **78** 1691, 2001.

[77] S.J. Potashnik, K.C. Ku, S.H. Chun, J.J. Berry, N. Samarth, and P. Schiffer. Effects of annealing time on defect-controlled ferromagnetism in $Ga_{1-x}Mn_xAs$. *Appl. Phys. Lett.*, **79** 1495, 2001.

[78] S. Sanvito and N.A. Hill. Influence of the local As antisite distribution on ferromagnetism in (Ga,Mn)As. *Appl. Phys. Lett.*, **78** 3493, 2001.

[79] S. Sanvito and N.A. Hill. Ab-initio transport theory for digital ferromagnetic heterostructures. *Phys. Rev. Lett.*, **87** 267202, 2001.

[80] G.A. Medvedkin, T. Ishibashi, T. Nishi, K. Hayata, Y. Hasegawa, and K. Sato. Room temperature ferromagnetism in novel diluted magnetic semiconductor CdMnGeP. *Jpn. J. Appl. Phys.*, **39** L949, 2000.

[81] Y. Matsumoto, M. Murakami, T. Shono, T. Hasegawa, T. Fukumura, M. Kawasaki, P. Ahmet, T. Chikyow, S. Koshihara, and H. Koinuma. Room temperature ferromagnetism in transparent transition metal-doped titanium dioxide. *Science*, **291** 854, 2001.

[82] K. Ueda, H. Tabata, and T. Kawai. Magnetic and electric properties of transition-metal-doped ZnO films. *Appl. Phys. Lett.*, **79** 988, 2001.

[83] W.R.L. Lambrecht, B. Segall, A.G. Petukhov, R. Bogaerts, and F. Herlach. Spin–orbit effects on the band structure and Fermi surface of ErAs and $Er_xSc_{1-x}As$. *Phys. Rev. B*, **55** 9239, 1997.

[84] D.R. Schmidt, A.G. Petukhov, M. Foygel, J.P. Ibbetson, and S.J. Allen. Fluctuation controlled hopping of bound magnetic polarons in ErAs:GaAs nanocomposites. *Phys. Rev. Lett.*, **82** 823, 1999.

Index